ELECTROCULTUE GARDENING

A COMPREHENSIVE STEP-BY-STEP MANUAL FOR BEGINNERS TO SUPERCHARGE ORGANIC VEGETABLE YIELDS WITH COPPER ANTENNA WIRES

BY

KEVIN KEATON

Copyright © 2024 by Kevin Keaton

All rights reserved. Without the publisher's prior written consent, no portion of this publication may be copied, distributed, or transmitted in any way, including by photocopying, recording, or other mechanical or electronic means, with the exception of brief quotations used in critical reviews and other noncommercial uses allowed by copyright law. To seek permission, send an email to the publisher at the following address, with the subject line "Attention: Permissions Coordinator."

DISCLAIMER

This book is intended for informational purposes only. Every attempt has been made by the author and publisher to guarantee the authenticity and comprehensiveness of the material presented here. However, they do not guarantee that the techniques and methods discussed will be suitable for every individual situation or yield specific results. The practices described in this book are based on the author's interpretations and understanding of electrophilic culture and should not be taken as expert advice.

Readers are encouraged to consult with professional agronomists, electricians, or other qualified experts before implementing any of the technologies or practices discussed. The implementation of the material in this book may result in direct or indirect personal, financial, or economic damages or losses for which the author and publisher shall not be liable.

Additionally, this book does not substitute for professional advice on agricultural, electrical, or environmental regulations, which can vary widely by location. Compliance with all local, state, and federal laws and regulations regarding electrophilic gardening is the sole responsibility of the reader.

The case studies and examples provided are for illustrative purposes and should be adapted to specific local conditions by the reader. They do not imply a guarantee of similar results, as outcomes can vary based on numerous environmental, geographical, and biological factors.

FOREWORD

"Electroculture Gardening: A Comprehensive Step-by-Step Manual for Beginners to Supercharge Organic Vegetable Yields with Copper Antenna Wires," by Kevin Keaton, we embark on a journey into the fascinating world of electrophilic culture, a frontier of agricultural science that promises to redefine our traditional understanding of gardening and farming. This book is designed not only as a comprehensive guide but also as an exploration of how integrating electricity with plant cultivation can unlock potentials that may seem like something out of science fiction.

The genesis of this book lies in the curiosity to understand how something as ubiquitous as electricity can be harnessed to dramatically enhance plant growth and productivity. The idea that we could improve crop yields, reduce resource waste, and potentially transform barren landscapes into lush, productive gardens with the help of controlled electrical impulses is a compelling proposition for anyone concerned with the future of food security and sustainable agriculture.

Electrophilic culture, or the use of electrical stimulation to promote plant growth, is a field that, although still in its infancy, has already shown promising results. Researchers and practitioners across the globe have begun to unveil the mechanisms through which plants perceive and respond to electrical signals, offering insights into how these can be optimized to support robust growth.

This book is intended for a wide audience—from hobby gardeners to professional agronomists—and aims to provide a clear and accessible overview of the principles and practices of electrophilic culture. It discusses the science behind how plants interact with electrical stimuli, provides practical guidance on setting up your own electrophilic systems, and provides insights into the challenges and benefits of this innovative approach.

We encourage readers to be open-minded to the possibilities that lay ahead as we flip the pages. The future of electrophilic culture is not only about enhancing how we grow food but also about understanding the deeper electrical nature of life itself. This book by Kevin Keaton is a step towards that future, offering a glimpse of what might be possible when we combine the forces of nature with human ingenuity.

Special Bonus for Our Readers

Thank you for purchasing Electroculture Gardening: A Comprehensive Step-by-Step Manual for Beginners to Supercharge Organic Vegetable Yields with Copper Antenna Wires. To enhance your reading experience and provide additional value, we have prepared an exclusive bonus just for you!

Please scan the QR Code below or visit the following link to download your free bonus material, which includes supplementary guides, advanced tips, and exclusive content to help you further your gardening skills.

[Insert QR Code Here]

TABLE OF CONTENTS

INTRODUCTION 8
    Welcome to the World of Electroculture Gardening 8
    Understanding the Basics of Electroculture 9
    The Promise of Enhanced Organic Yields 10
    How to Use This Manual 11

1 13
THE SCIENCE BEHIND ELECTROCULTURE 13
    Understanding Electroculture Fundamentals 13
    Historical Context and Evolution 14
    The Role of Copper Antenna Wires in Plant Growth 15
    Understanding Plant Bioelectricity 17
    How Plants Use Electricity 18
    Scientific Research and Findings 20
    Debunking Myths and Misconceptions 21

2 24
GETTING STARTED WITH ELECTROCULTURE 24
    Preparing Your Garden for Electroculture 24
    Essential Tools and Materials 25
    Choosing the Right Copper Wires 27
    Soil Preparation and Testing 28
    Understanding Soil Types 29
    Testing Soil pH and Nutrients 31
    Amending Soil for Optimal Growth 32

3 34
INSTALLING COPPER ANTENNA WIRES 34
    Selecting the Perfect Antenna Design 34
    Step-by-Step Installation Guide 35
    Positioning Antennas for Maximum Effectiveness 37
    Integrating Antennas with Existing Structures 38
    Safety Considerations 39

4 ....**43**

TROUBLESHOOTING COMMON ISSUES..**43**

   Diagnosing and Addressing Electroculture Challenges 43

   Fine-Tuning Your Setup for Maximum Efficiency 44

   Sustainable Practices and Organic Solutions..46

5 ....**48**

THE FUTURE OF ELECTROCULTURE **48**

   Current Research and Emerging Trends in Electroculture 48

   Predictions and Innovations on the Horizon...49

CONCLUSION **52**

   Reflecting on Your Electroculture Journey 52

   The Future of Electroculture Gardening...53

   Encouragement and Final Tips..54

INTRODUCTION

Welcome to the World of Electroculture Gardening

Welcome to the world of Electroculture gardening, a transformative practice that intertwines the elegance of nature with the prowess of modern science to supercharge the growth and vitality of your garden. This guide is your gateway into an innovative method that harnesses the earth's natural electrical energies to enhance plant development, boost yields, and fortify the health of your soil. Embarking on this journey not only introduces you to a groundbreaking agricultural practice but also invites you to participate in a sustainable movement that aligns closely with the rhythms of the natural world.

Electroculture gardening is not just about planting seeds and watching them grow; it's about redefining the interaction between nature and technology. For centuries, humans have cultivated the land, relying on water, soil, and sunlight to nurture their crops. However, the integration of Electroculture into gardening practices marks a significant leap forward, offering a method that uses electrical currents to stimulate plant growth in ways that traditional methods cannot match. This technique taps into the subtle yet powerful electrical fields that permeate our natural environment, using them to enhance the biological processes that drive plant growth.

At its core, Electroculture is about more than just agriculture; it's about a philosophy that values efficiency, sustainability, and harmony with nature. It challenges conventional gardening paradigms by introducing an element that many may find futuristic yet deeply rooted in scientific research. The principle is simple yet profound: by enhancing the natural bioelectric environment around plants, we can significantly increase their ability to absorb nutrients, process them more efficiently, and improve overall plant resilience against diseases and pests.

The journey into Electroculture starts with an understanding of how electrical energy interacts with plant life. It's a fascinating realm of study where biology meets physics, revealing that all living organisms, including plants, respond to electrical stimuli. This responsiveness can be leveraged to promote faster growth, enhance yield, and even extend the growing season beyond traditional limits. For gardeners, this means the ability to achieve more with less: less water, less fertilizer, and fewer pesticides, thus reducing the environmental footprint of their gardening activities.

Additionally, Electroculture extends an invitation to gardeners to join a forward-thinking group that is expanding the realm of what is feasible in horticulture. It's a practice that encourages experimentation and observation, requiring an inquisitive mind and a willingness to engage with the garden as a dynamic ecosystem. As you go through this handbook more thoroughly, you will discover not only how to set up an Electroculture system technically but also some of the scientific ideas behind its operation.

The transformative potential of Electroculture extends beyond individual gardens. It represents a forward-thinking approach to food production that can be scaled up to meet community or commercial needs without compromising the integrity of the natural environment. It offers a solution to some of the pressing challenges of modern agriculture, such as the need to increase production in the face of diminishing natural resources and the growing demand for organic and sustainably grown produce.

This manual is designed to guide you step-by-step through the process of setting up and maintaining an Electroculture garden. The material offered here will give you the knowledge and resources you need to start this green adventure, from comprehending the fundamentals of how and why it works to fixing common issues and investigating the future potential of this exciting subject. Whether you are a hobbyist looking to experiment with a new gardening technique or a seasoned farmer seeking sustainable solutions, Electroculture offers a compelling pathway to enhancing your gardening practices while staying aligned with ecological principles.

As we turn each page, remember that Electroculture is not just about technology; it's about enhancing our connection with the earth and contributing to a sustainable future. It's a journey that promises to enrich your understanding of the natural world and transform your approach to gardening. Together, let's go on this life-changing journey that will transform knowledge into action and turn your garden into a living example of what's possible when we align with the planet's natural energies.

Understanding the Basics of Electroculture

Understanding the basics of Electroculture is essential for anyone interested in harnessing its capabilities to supercharge their garden. This innovative approach to gardening leverages the Earth's natural electrical energies, which play a significant role in enhancing plant growth, vitality, and yield. At its core, Electroculture is a method that merges the timeless art of gardening with the nuanced science of electricity, bringing a new dimension to nurturing plants.

Electroculture operates on the premise that the Earth's atmosphere and soil are alive with electrical charges that can influence plant growth. These charges are part of a natural cycle, influenced by factors such as solar radiation, lightning, and the Earth's magnetic field. These natural energies are omnipresent, subtly influencing biological processes. In the realm of Electroculture, we take advantage of this natural phenomenon by amplifying these electrical charges in a controlled manner to directly benefit the plants.

The basic mechanism through which Electroculture operates involves introducing additional electrical fields into the garden's ecosystem. This is typically achieved using conductive materials like copper wires, which are installed around or within the plant beds. These wires are not just passive structures; they actively channel and direct the Earth's natural electrical energies in a way that plants can benefit from more directly. When these wires are charged, they create an electrical field that permeates the soil and interacts with the natural bioelectricity of plant cells.

This interaction stimulates the plants in several beneficial ways. Firstly, it enhances nutrient uptake from the soil, as the electrical fields can help dissolve and transport nutrients to the roots more efficiently. This process is akin to giving the plants a continuous, low-dose nutrient boost, which can lead to more robust growth and enhanced productivity. Moreover, these electric fields stimulate root growth, leading to a stronger and more expansive root system. With more extensive root systems, plants can access a larger volume of soil and, thus, more resources, further supporting their growth.

Electroculture also influences the water efficiency of plants. Electrical stimulation improves how well plants use water, lowering the need for frequent watering and promoting plant growth even in arid environments. This aspect of Electroculture is particularly valuable in areas facing water scarcity or those looking to reduce their water usage without sacrificing plant health or crop yields.

Beyond enhancing growth and resource efficiency, Electroculture can contribute to plant health by bolstering their immune systems. The electrical fields can induce a kind of mild stress response in plants, which activates their natural defense mechanisms. This may lessen their susceptibility to illnesses and pests, thereby lowering the need for chemical treatments like fungicides and insecticides.

Understanding Electroculture also involves appreciating its subtle nature. Unlike more intrusive agricultural technologies, Electroculture works in harmony with nature. It does not alter the genetic makeup of the plants, nor does it rely on chemical inputs. Instead, it simply enhances the natural environmental factors that plants already use to grow. This makes Electroculture an exceptionally sustainable practice that aligns with organic gardening principles and ecological conservation.

For gardeners and farmers, adopting Electroculture means stepping into a world where science and nature coexist in closer harmony. It offers a way to significantly boost garden and crop productivity while maintaining an organic and environmentally friendly approach. This straightforward yet effective technique creates new opportunities for raising food output without endangering the planet's health or the food's nutritional content.

The Promise of Enhanced Organic Yields

The adoption of Electroculture in gardening and agriculture promises a revolution in the way we cultivate our plants, emphasizing sustainability and efficiency. This method is particularly compelling for its potential to enhance organic yields, which is a crucial consideration in an era where both environmental sustainability and food security are of paramount concern. The promise of Electroculture lies in its ability to increase crop yields, improve plant health, and enhance soil quality without relying on chemical inputs, thus supporting an organic farming ethos.

One of the most compelling advantages of Electroculture is the significant increase in crop yields. By optimizing the electrical environment around plants, Electroculture stimulates their natural growth processes more effectively than traditional farming methods. The electrical fields generated around the plants help in more efficient nutrient absorption from the soil, enhanced photosynthesis, and quicker growth cycles. This stimulation leads to healthier plants that grow faster and produce more. For farmers and gardeners, this means higher productivity per square foot of garden or field, which is invaluable in maximizing the output of limited space—a common challenge in urban and small-scale farming environments.

Moreover, Electroculture impact on plant health is profound. Plants grown under Electroculture conditions exhibit enhanced resilience against common diseases and pests. This resilience is partly due to the strengthened immune response stimulated by the electrical fields, which helps plants naturally fend off diseases without the need for chemical fungicides or pesticides. Healthier plants not only yield more but also have a better quality of produce, which is particularly important for organic markets where the quality often supersedes quantity in value. Additionally, healthier plants can better withstand adverse weather conditions, making Electroculture a valuable tool in areas prone to environmental stresses that would otherwise compromise crop success.

Another significant benefit of Electroculture is its positive impact on soil quality. Traditional agricultural practices often degrade soil health over time through the excessive use of chemical fertilizers and soil tilling. In contrast, Electroculture contributes to soil vitality without chemical inputs. The technique enhances the natural bioactivity in the soil, including promoting beneficial microorganisms essential for organic soil health. These microorganisms play a critical role in decomposing organic matter, cycling nutrients, and naturally aerating the soil, all of which contribute to the long-term sustainability of the soil.

Electroculture also encourages better water management, a critical aspect of maintaining soil quality and overall agricultural sustainability. The method enhances the plant's ability to absorb and utilize water efficiently, thereby reducing runoff and minimizing the need for frequent watering. This not only conserves water—a precious resource in many farming regions—but also prevents the leaching of nutrients and the erosion of topsoil, further maintaining soil integrity and fertility.

Furthermore, Electroculture aligns perfectly with the principles of organic farming, which emphasize ecological balance, biodiversity, and cycles adapted to local conditions without the use of synthetic pesticides or fertilizers. By enhancing plant growth through natural means, Electroculture allows organic farmers to achieve higher yields while maintaining the integrity of their organic certification and the purity of their produce.

The broader adoption of Electroculture could transform agricultural practices by providing a method that increases productivity while also preserving and enhancing the environmental resources upon which farming depends. It offers a path towards a sustainable agricultural future where technology and nature work hand in hand to feed the growing global population without depleting the Earth's natural resources.

How to Use This Manual

This manual is designed to be both a comprehensive guide and a practical tool for anyone interested in integrating Electroculture into their gardening or farming practices. Regardless of your level of experience with gardening or familiarity with the idea of utilizing electrical energies to promote plant growth, the knowledge and methods offered here will provide you the tools you need to carry out Electroculture successfully. Here's how to get the most out of this guide so you may successfully use these creative methods in a variety of garden configurations.

The manual is structured to take you step-by-step through the essential concepts, setup procedures, and maintenance tips for Electroculture. It starts with fundamental theories, moves on to practical applications, and covers troubleshooting and advanced tips, making it suitable for both beginners and those looking to deepen their understanding of Electroculture.

The manual is divided into distinct sections, each aimed at building your understanding and skills in a logical progression. Initially, you'll find chapters focused on the basics—what Electroculture is, how it works, and the scientific principles behind it. This foundational knowledge is crucial as it prepares you for more complex topics and practical applications that follow.

Each chapter is designed to address different aspects of Electroculture, from soil preparation and wire installation to system maintenance and optimization. Within these chapters, practical steps, illustrated diagrams, and specific instructions are provided to help you apply what you've learned. For example, the section on installing copper antenna wires offers detailed guidance on choosing the right materials, measuring and cutting wire, and correctly positioning the wires in your garden for maximum effect.

To ensure that you can adapt the instructions to your specific situation, the manual includes various scenarios and garden setups—from small backyard gardens to larger agricultural plots. This approach helps you visualize how Electroculture can be scaled and adapted to meet different needs and conditions.

The manual also incorporates interactive elements such as troubleshooting guides and FAQ sections. These elements are designed to engage you actively in solving common problems you might encounter and to provide quick answers to frequently asked questions. By interacting with these elements, you gain a deeper understanding of how to fine-tune your Electroculture system and enhance its performance over time.

Beyond technical instructions, the manual provides practical tips and real-world applications of Electroculture. This includes advice on integrating Electroculture with other gardening practices, such as organic farming or permaculture, and suggestions for modifying your approach based on the specific crops you are growing or the climatic conditions of your area.

Finally, the manual encourages continuous learning and adaptation. Electroculture, like any innovative technology, evolves over time as new research emerges and as practitioners share their experiences. To stay updated, you are encouraged to engage with the broader Electroculture community, participate in forums, and attend workshops or seminars. This manual should be seen not just as a static resource but as a starting point for an ongoing journey in sustainable agriculture.

1

THE SCIENCE BEHIND ELECTROCULTURE

Understanding Electroculture Fundamentals

Understanding the fundamentals of Electroculture involves delving into the scientific principles that underpin this innovative agricultural practice. At its core, Electroculture is about utilizing the Earth's natural electrical energies to stimulate plant growth, a concept rooted in the intersection of botany, physics, and environmental science. This exploration not only reveals how Electroculture works but also why it is so effective in enhancing plant vitality and productivity.

Electroculture operates on the basic scientific premise that all living organisms, including plants, interact with their environment through electrical signals. The Earth itself is a dynamic source of electrical energy, with natural currents and fields generated by various natural phenomena such as solar radiation, atmospheric changes, and geomagnetic forces. These environmental and electrical forces are subtle yet influential, impacting biological processes at the cellular level in plants.

Plants naturally produce and respond to electrical signals. These bioelectrical signals are integral to plant physiology, influencing processes such as photosynthesis, nutrient uptake, and growth regulation. Electroculture techniques amplify these natural electrical interactions by introducing additional external electrical stimuli, which enhance the plants' existing bioelectrical activities.

The introduction of controlled electrical fields around plants facilitates a more efficient movement of ions and nutrients in the soil. This enhanced ion mobility increases nutrient availability to the plant roots, which can absorb them more readily. Furthermore, the electric fields can help regulate the opening and closing of stomata on plant leaves, which controls transpiration and gas exchange, optimizing the plant's respiratory processes and water usage.

One of the key components of Electroculture is the use of conductive materials, such as copper wires, strategically placed in and around the soil to create an induced electrical field. When these wires are charged, they interact with the natural electrical field of the Earth, enhancing it locally around the plants.

This interaction stimulates the plants' natural growth processes by mimicking and intensifying the beneficial electrical stimuli that plants would normally experience in an optimal natural environment.

This electrophysical stimulation has been shown to accelerate growth, enhance root development, and improve overall plant health. Research indicates that plants growing in electrically enhanced environments tend to have greater biomass, higher photosynthetic efficiency, and better resistance to diseases and pests compared to those grown under normal conditions.

The effectiveness of Electroculture has been supported by numerous scientific studies and practical observations. For instance, experiments have demonstrated that seeds exposed to an electrical field germinate faster and produce seedlings with a higher survival rate. Similarly, in agricultural settings, crops grown with Electroculture techniques often show increased yields and better-quality produce.

Moreover, the science behind Electroculture also touches on the ecological impacts of using electrical stimulation in agriculture. Unlike chemical enhancers, electrical enhancements do not contribute to soil degradation or pollution. Instead, they offer a sustainable alternative to boost plant growth and productivity without adverse environmental impacts.

The principles of Electroculture are not only relevant to increasing agricultural output but also have broader implications for understanding how plants interact with their environment. This knowledge contributes to other fields, such as ecological management, where understanding the electrical aspects of plant-environment interactions can help in the restoration of ecosystems and the management of natural resources.

Historical Context and Evolution

Electroculture, while seemingly a product of modern scientific advancements, has roots that stretch far back into history, intertwining with the agricultural practices of ancient civilizations. This journey from past to present shows how the understanding and application of electrical principles in agriculture have evolved, leading to sophisticated techniques that harness the Earth's natural electrical forces to enhance plant growth.

The concept of using electrical energy to influence plant vitality can be traced back to the 1740s when Abbot Nollet, a French clergyman and scientist, first experimented with electrostatic effects on plant growth. Nollet's curiosity about the natural world and its unseen forces led him to expose seeds and plants to static electrical charges, noting accelerated growth rates compared to those not subjected to such conditions. These early experiments sparked interest among European scientists, setting the stage for further exploration into the relationship between electricity and plant biology.

By the 19th century, the field of Electroculture had gained momentum with the work of Sir Humphry Davy, an English chemist and inventor who also delved into the effects of electrical currents on vegetation. His experiments provided a more scientific foundation for the idea that electrical treatments could modify plant growth processes. Davy's work inspired others across Europe to conduct their experiments, many of which demonstrated that electricity could indeed enhance seed germination and increase crop yields.

As the scientific community in the 19th century continued to expand their understanding of electricity, farmers, and agriculturalists began to see practical applications for these discoveries. The invention of the voltaic pile by Alessandro Volta, which allowed for a continuous source of electrical energy, opened new doors for applying electrical science in agriculture. Experiments during this period often involved running currents through sections of soil or directly into plant systems, observing increases in growth and health that seemed to confirm the efficacy of electrical intervention.

In the early 20th century, the interest in Electroculture was such that it was found in support in various agricultural stations and universities, which dedicated resources to the study of electrical effects on plant systems. Reports from this era documented substantial improvements in yield and plant robustness when crops were grown under conditions that included electrical stimulation. Researchers theorized that electricity might influence a range of biological processes, from nutrient uptake to enzymatic activity, thereby offering a potent tool for agricultural enhancement.

However, despite these promising results, Electroculture has not become mainstream. The reasons were manifold, including the inconsistency of results, the onset of the chemical fertilizer boom, and the Great Wars, which shifted scientific focus to other pressing concerns. The widespread adoption of chemical solutions in agriculture, which promised more immediate and visually apparent results, overshadowed the subtler and less understood benefits of Electroculture.

The resurgence of interest in Electroculture that we witness today is largely due to the increasing environmental concerns related to chemicals, agricultural practices, and the sustainable agriculture movement. As the negative impacts of pesticides and synthetic fertilizers on ecosystems and human health have become more apparent, there is renewed interest in finding more harmonious ways to boost crop productivity. Modern Electroculture, equipped with better technology and a deeper understanding of electrical phenomena in biological systems, stands out as a promising alternative.

Modern Electroculture techniques now incorporate sophisticated technology such as solar-powered systems to generate electrical fields, precise monitoring equipment to regulate the intensity and frequency of electrical inputs, and advanced materials science to develop better conductors and electrodes. These improvements have made Electroculture more accessible and applicable on a larger scale, promising not only increased agricultural yields but also enhanced sustainability.

The evolution of Electroculture is a testament to the persistent human curiosity about natural forces and their potential applications. From the static electricity experiments of the 18th century to the solar-powered Electroculture systems of today, this field has matured into a viable scientific discipline that offers a blend of ancient wisdom and modern technology. This blend not only aims to increase food production but also to do so in a way that respects and preserves the natural environment, making Electroculture a critical component of the future of sustainable agriculture.

The Role of Copper Antenna Wires in Plant Growth

In the realm of Electroculture, copper antenna wires play a pivotal role in enhancing plant growth by harnessing and directing the Earth's natural electrical energies into the soil and plants themselves. These wires, when properly installed and managed, create an environment that is conducive to accelerated growth rates, improved health, and increased yields. To understand how copper wires function within an

Electroculture system, it is essential to explore their properties, the mechanics of their operation, and the outcomes they facilitate.

Copper is chosen for its excellent electrical conductivity—surpassed only by silver among commercial metals—which allows it to efficiently transfer electric charges. Additionally, copper is relatively resistant to corrosion and degradation in soil environments, which makes it durable and suitable for prolonged use in Electroculture setups. These properties make copper an ideal material for creating antenna wires that can effectively conduct and distribute electrical energy in a garden or farm setting.

The basic premise behind using copper wires in Electroculture is to enhance the natural bioelectric environment around the plants. This is achieved by introducing a low-intensity electric field into the area where crops are growing. The wires are typically buried a few inches below the soil surface and are arranged in specific patterns—often grid-like—to ensure even distribution of the electrical charge. When an electrical current is passed through these wires, it generates an electric field that permeates the surrounding soil and interacts with the plant roots.

This interaction between the electric field and the plants is multifaceted. Firstly, the electric field increases the mobility of ions in the soil. This is crucial because plants absorb nutrients in their ionic form. Elements like nitrogen, phosphorus, potassium, and trace minerals become more available to plant roots when they are ionized and mobile. As a result, plants can uptake nutrients more efficiently and in larger quantities, which directly translates to better growth, greater biomass, and enhanced metabolic functions.

Moreover, the electric field influences the water molecules in the soil. Water is a polar molecule, meaning it has a slight positive charge on one side and a negative charge on the other. The electric field generated by the copper wires helps to align these water molecules in a way that makes water uptake easier for plants. This not only improves hydration but also enhances the plant's ability to absorb soluble nutrients dissolved in water.

Electrical stimulation from copper wires also affects the physiological processes of the plants directly. It has been observed that exposure to certain electric fields can enhance photosynthesis, the process by which plants convert light energy into chemical energy. Increased photosynthesis means more energy is available for growth and fruiting, which is ultimately what farmers and gardeners aim to maximize.

Another significant impact of using copper wires in Electroculture is on plant health. The electric fields can stimulate a plant's natural defense mechanisms, making them more resistant to diseases and pests. This is akin to how low-dose stressors can trigger organisms to develop resilience through a process known as hormesis. Plants exposed to these mild electrical stimuli often exhibit stronger immune responses, which reduce the need for chemical pesticides.

Furthermore, the structured setup of copper wires in the soil can also help to improve soil structure over time. The gentle warming effect of the passing current can promote beneficial microbial activity in the soil. These microbes play a critical role in organic matter decomposition, nutrient cycling, and maintaining soil health. Healthy soil is fundamental for sustainable agricultural practices, as it supports robust plant growth and maintains ecological balance.

In practical terms, setting up a copper wire system for Electroculture involves careful consideration of the specific crop requirements, soil conditions, and climatic factors. The wires must be positioned at a depth and spacing that optimizes their interaction with the plant roots and soil type. They must also be connected to a safe and reliable power source that can deliver a consistent and appropriate level of electrical charge.

The implementation of copper wires in an Electroculture system represents a harmonious blend of ancient understanding and modern technological application. By leveraging the conductive properties of copper, farmers, and gardeners can create an energetically enhanced environment that promotes vigorous plant growth and sustainable agricultural productivity. This innovative use of copper antenna wires not only underscores the potential of Electroculture as a transformative agricultural practice but also highlights the importance of integrating scientific knowledge with natural processes to meet the challenges of modern food production.

Understanding Plant Bioelectricity

Plants, much like humans and animals, exhibit a fascinating aspect of physiology known as bioelectricity. This refers to the electrical signals that plants generate and respond to as part of their biological processes. Understanding plant bioelectricity is essential in the context of Electroculture, as it provides insights into how electrical interventions can be harnessed to enhance plant health and productivity.

Bioelectricity in plants is an integral part of their survival and adaptation mechanisms. It involves the movement of ions across plant cell membranes, which creates electrical potentials. These potentials are crucial for various physiological processes, including photosynthesis, nutrient uptake, and response to environmental stressors.

The primary components of plant bioelectricity are the ion channels and pumps embedded in the plant cell membranes. These channels and pumps regulate the flow of ions, such as potassium, calcium, and chloride, into and out of the cells. The movement of these ions is what generates bioelectrical signals. For example, when a plant encounters a drought condition, it might close its stomata to prevent water loss. The signal to close the stomata is facilitated by changes in the electrical potential across the cell membranes of the guard cells, which are controlled by the dynamic flux of ions.

This electrical signaling system is not just limited to individual cells or parts of a plant but can propagate throughout the entire organism. When one part of a plant is wounded, an electrical signal can quickly spread to other parts, initiating systemic responses such as the production of defensive chemicals or the strengthening of cell walls. This rapid communication within the plant enhances its ability to respond to environmental challenges efficiently.

Moreover, plant bioelectricity is also involved in longer-term adaptive responses and growth patterns. For instance, the direction and rate of root growth can be influenced by electrical gradients in the soil, which are perceived by the root tips. These gradients help the roots navigate toward water and nutrient sources, optimizing the plant's chances of survival and growth.

The natural bioelectric activities of plants are influenced by their environment. External factors such as light, temperature, soil quality, and water availability can affect the bioelectric potentials of plants. Understanding these interactions is crucial for optimizing agricultural practices and can provide a basis for the use of Electroculture.

In the context of Electroculture, the enhancement of natural plant bioelectricity through external electrical stimulation—such as with copper wires or other conductive systems—can further optimize these physiological processes. By applying a mild external electric field, we can influence the ion fluxes and potentials in a way that promotes more vigorous growth, improves nutrient efficiency, and enhances the plant's natural defense mechanisms.

This interaction between external electrical fields and plant bioelectricity is believed to stimulate various growth-related processes. For example, enhanced bioelectric signaling can lead to more efficient photosynthesis, better nutrient and water uptake, and quicker response to environmental stresses. These benefits are reflected in healthier, faster-growing plants with improved yields.

Moreover, the application of Electroculture does not alter the genetic makeup of the plants but rather works with the natural bioelectrical systems that plants already use. This makes Electroculture a highly sustainable agricultural technique that complements organic farming methods, avoiding the use of genetic modification or chemical enhancers.

How Plants Use Electricity

Plants inherently use electricity as a natural part of their physiological processes, though this is often on a subtle scale. Bioelectricity in plants involves the movement of ions across cell membranes, creating electric potentials that signal various biological functions necessary for growth and survival. Understanding how plants use this natural bioelectricity provides a foundation for exploring how added electrical inputs, such as those used in Electroculture, can further enhance these processes, leading to improved growth rates, increased productivity, and better overall plant health.

The interaction between plant bioelectricity and added electrical inputs starts at the cellular level. Plant cells, like animal cells, have electrically charged membranes created by ion gradients. These gradients are maintained by ion pumps and channels that move ions in and out of the cell. The resulting voltage difference across the cell membrane is what drives the electrical signaling in plants.

Electrical Signaling in Plants

Plants use this electrical signaling primarily for internal communication. It plays a crucial role in coordinating responses to environmental stimuli. For example, when a plant part is wounded, an electrical signal is generated at the site of injury. This signal, a change in the electrical potential, travels to other parts of the plant, triggering defensive mechanisms such as the production of secondary metabolites that deter herbivores or the fortification of cell walls to prevent pathogen invasion.

This natural bioelectric signaling is also critical in managing physiological processes such as photosynthesis and transpiration. For instance, the opening and closing of stomata—the pores on leaves through which gas exchange occurs—are regulated by changes in electrical potentials triggered by environmental cues like light intensity and humidity.

Enhancing Natural Electricity with External Inputs

When external electrical inputs are introduced to plants via techniques like Electroculture, they essentially amplify the natural bioelectric environment around the plant roots and throughout the plant structure. This can be particularly effective because it enhances the plant's existing capabilities in nutrient uptake and growth regulation.

Enhanced Nutrient Absorption: Electrical fields influence the mobility of charged nutrient ions in the soil. By applying a mild external electric field, ions such as nitrogen, phosphorus, potassium, and trace elements can be driven toward the root surfaces more effectively. This increases the concentration gradient, which enhances passive and active absorption of these nutrients by plant roots. The more efficient uptake translates into better nutrition, which fuels more robust growth and development.

Stimulation of Root Growth: The growth of plant roots is also influenced by electric fields. Roots naturally grow toward positive charges (anions) in the soil, a phenomenon known as electrotropism. By manipulating the electrical conditions in the soil, it's possible to guide root growth, encouraging roots to expand in nutrient-rich areas or areas where water is more accessible, thus optimizing the plant's resource acquisition.

Acceleration of Photosynthesis: External electrical stimulation can enhance photosynthetic activity by influencing the opening and closing of stomata, thereby optimizing the plant's gas exchange processes. Efficient gas exchange ensures that carbon dioxide levels within the leaf are optimal for photosynthesis while also managing water use efficiency, especially under conditions of water stress.

Increased Disease Resistance: The electrical stimulation can enhance the plant's immune response, making it quicker and more robust. This can result in heightened resistance to diseases and pests, reducing the need for chemical interventions. Electrical signals can trigger the production of phytochemicals that serve as deterrents or toxins to pests and pathogens.

The interaction between added electrical inputs and plant bioelectricity in Electroculture is a vivid example of how technology can mimic and enhance natural processes for agricultural benefit. By understanding and leveraging these interactions, farmers and gardeners can use Electroculture to achieve higher yields, more sustainable practices, and healthier plants. This synergy between technology and biology not only holds the promise of more productive agriculture but also aligns with eco-friendly practices that support long-term agricultural sustainability.

Scientific Research and Findings

The field of Electroculture, which harnesses electrical stimulation to enhance plant growth and productivity, is supported by a body of scientific research and empirical findings that validate its benefits. Over the years, numerous studies have been conducted that demonstrate the positive effects of electric fields on various aspects of plant biology, from germination and growth to resistance against pests and diseases. These findings not only offer a better understanding of the mechanisms at play but also provide a strong argument for the integration of Electroculture techniques in modern agriculture.

One of the earliest scientific explorations into the effects of electricity on plant growth dates back to the 18th century with the experiments conducted by Luigi Galvani and later by his contemporary, Alessandro Volta. Both scientists discovered that electric currents could influence the movement of plant tissues and fluids. Fast forward to the 20th century, when the field gained renewed interest, researchers began to systematically investigate the role of electric fields in promoting plant growth and protecting against environmental stresses.

A significant study conducted in the 1970s at the University of California, Berkeley, found that seeds exposed to an electric field had a higher germination rate compared to those that were not. The researchers theorized that the electric field might affect the permeability of the seed membrane, thus enhancing water uptake and activating enzymes that trigger germination. This study paved the way for further research into how electric fields could be used to improve seed germination across various plant species.

Further research expanded into the growth phases of plants, with numerous studies throughout the 1980s and 1990s demonstrating that low-intensity electric fields could stimulate growth in plant shoots and roots. One such study measured increased growth rates in tomato plants exposed to static electric fields. The findings suggested that the electric field enhanced the activity of growth-promoting hormones, leading to faster and more robust plant development. This was a crucial discovery, showing that Electroculture could be a viable technique not only for improving yields but also for enhancing the speed at which plants mature.

The impact of Electroculture on nutrient uptake has also been the subject of extensive study. Research conducted in the early 2000s demonstrated that electrically charged soils could alter the availability of nutrients by changing the soil pH and the solubility of minerals. This effect facilitated a more efficient uptake of essential nutrients such as nitrogen, phosphorus, and potassium, which are critical for plant growth. Plants grown in electrically enhanced soil conditions showed higher biomass and yield, attributed to improved nutrient availability and uptake efficiency.

Electroculture benefits extend beyond growth stimulation to include enhanced resistance to pests and diseases. A landmark study published in the late 1990s showed that plants exposed to specific patterns of electric pulses could increase their production of defensive chemicals, such as phenolics and phytoalexins, which protect against microbial infections and insect attacks. This adaptive response reduced the need for chemical pesticides, aligning Electroculture with sustainable farming practices that seek to minimize chemical inputs.

The technique has also been found to be beneficial in improving plant stress responses to environmental challenges such as drought and salinity. Studies have shown that electrical stimulation can enhance the synthesis of stress hormones like abscisic acid, which helps plants conserve water during drought conditions. Additionally, exposure to electric fields has been observed to help plants manage the toxic effects of saline soils, possibly by affecting ion transport mechanisms in a way that prevents the accumulation of harmful salts within plant tissues.

Electroculture has not only been explored in field trials but also in controlled environments like greenhouses, where researchers have successfully used electric fields to control the growth conditions and study the resulting effects on plant health and productivity. These studies have consistently shown that even slight alterations in the electrical environment can have significant positive effects on plant growth and resilience.

Moreover, the scalability of Electroculture from laboratory settings to real-world agricultural applications has been demonstrated in various pilot projects around the world. In regions where traditional agriculture is challenged by poor soil quality or adverse climate conditions, Electroculture has provided a means to boost productivity without the extensive use of chemical fertilizers or irrigation, thus offering a cost-effective and environmentally friendly alternative.

Debunking Myths and Misconceptions

Electroculture, despite its scientific backing and proven benefits, is surrounded by various myths and misconceptions that can obscure its understanding and adoption. These misconceptions often stem from a lack of awareness about how Electroculture works, its safety, and its implications for both the environment and agricultural productivity. Addressing these myths is crucial for advancing the practical applications of Electroculture and ensuring it is viewed within the correct scientific context.

One common myth is that Electroculture involves dangerous levels of electricity that could harm the plants, the environment, or the farmer. This misconception likely arises from a misunderstanding of the types of electrical currents used in Electroculture. The reality is that Electroculture typically uses very low-voltage and low-amperage electric currents, which are not only safe for the plants but also for humans handling the installations. These electrical inputs are carefully controlled and much lower than the thresholds that could cause harm, ensuring they enhance growth without damaging cellular structures.

Another widespread misconception is that Electroculture is unnatural and could somehow disrupt the ecological balance. Critics often group Electroculture with other agricultural interventions like genetic modification or chemical usage, which are seen as intrusive or manipulative. However, Electroculture fundamentally differs in that it leverages a natural environmental factor—electricity—which is already present in nature. Plants have evolved to respond to natural electrical fields, such as those generated during thunderstorms. Electroculture simply amplifies these existing environmental cues in a controlled manner, enhancing their beneficial effects without introducing foreign chemicals or genes into the ecosystem.

There is also skepticism regarding the efficacy of Electroculture, with some doubting whether it can truly improve plant growth and yield. This skepticism is often fueled by anecdotal reports of unsuccessful attempts to use Electroculture, which may not have adhered to the scientifically established protocols. Numerous peer-reviewed studies have documented the positive effects of Electroculture on a variety of plants across different stages of growth—from seed germination to increased biomass and yield. These studies provide a robust evidence base that demonstrates Electroculture effectiveness when applied correctly, following the specific electrical parameters suitable for each plant species and growth condition.

The belief that Electroculture is too complex and impractical for widespread use is another misconception that hinders its adoption. While the technology requires some initial setup and understanding of the basic principles, modern advancements have made Electroculture systems increasingly user-friendly and accessible. Innovations in automated controls, portable electric field generators, and solar-powered systems have simplified the integration of Electroculture into existing agricultural practices. These developments make it easier for farmers to implement Electroculture without needing extensive technical expertise, thereby broadening its applicability and potential for impact.

Finally, there is a misconception that Electroculture is a standalone solution that should replace other agricultural practices. In reality, Electroculture is best used as a complementary technique alongside other sustainable practices such as crop rotation, organic fertilization, and integrated pest management. The integration of Electroculture can enhance practices, leading to a holistic approach to sustainable agriculture that maximizes productivity while minimizing environmental impact.

Debunking these myths and misconceptions is essential for fostering a realistic understanding of Electroculture. By clarifying its scientific basis and practical applications and by demonstrating how it safely and effectively integrates into broader agricultural systems, Electroculture can be recognized as a valuable tool in the sustainable agriculture toolkit. This recognition will encourage more widespread adoption and innovation in the field, ultimately contributing to more productive and sustainable agricultural practices worldwide.

Thank You for Starting This Journey!

We hope you've enjoyed the introduction to **Electroculture Gardening**. If you're finding the information helpful, we would greatly appreciate your taking a moment to leave a review on Amazon. Your feedback not only helps us improve but also assists other readers in understanding the benefits of this book.

How You Can Share Your Review:

Through Amazon.com:

- Go to the Amazon page where you found my book.
- Navigate to the 'Customer Reviews' section.
- Click on 'Write a customer review' to share your valuable insights.
- Instant QR Code Access: Simply scan the QR code below with your smartphone to be directed to the Amazon review section.

2

GETTING STARTED WITH ELECTROCULTURE

Preparing Your Garden for Electroculture

Integrating Electroculture into your garden begins with careful assessment and preparation of your garden space. This foundational step ensures that the transition to using electrical enhancements is both smooth and effective, paving the way for improved plant growth and yield. The preparation process involves a series of thoughtful evaluations and modifications to the garden environment to accommodate the specific needs and mechanics of Electroculture.

The first step in preparing your garden for Electroculture is to conduct a thorough assessment of the current soil conditions. Soil quality is paramount in Electroculture as it affects how well the electric fields can be conducted and how effectively the plants can utilize the enhanced conditions. Testing the soil for its pH level, electrical conductivity, moisture content, and nutrient profile provides a baseline understanding of what might need to be adjusted. For instance, soils with very low conductivity might require the addition of certain amendments to improve ion flow, which is crucial for the effective transmission of electrical currents.

After assessing the soil, the next step is to consider the layout and structure of your garden. Electroculture systems often require specific configurations of conductive materials, such as copper wires, to create an effective electric field. Planning the layout involves determining where and how these wires will be installed in relation to the plant rows or beds. It's important to ensure that the wires are placed at a depth and spacing that optimizes their interaction with the root systems of the crops being grown. This might involve some trial and error, as well as adjustments based on the specific growth habits and root depths of different plant species.

Water management is another crucial aspect to consider when preparing for Electroculture. Because electrical conductivity is influenced by moisture levels in the soil, maintaining optimal moisture is essential. This doesn't just mean ensuring the garden is regularly watered; it also involves setting up an irrigation system that can keep the soil moisture at a consistent level without overwatering or causing drainage issues. In some cases, integrating drip irrigation systems that deliver water directly to the root zones of plants can help maintain the necessary balance of moisture for effective Electroculture.

Additionally, the electrical infrastructure itself must be carefully planned and safely installed. This includes setting up a reliable and safe source of electricity that can power the system without posing a hazard. Solar-powered systems are often favored in Electroculture setups for their sustainability and efficiency, but whatever power source is chosen, it must be capable of delivering a consistent and controlled electrical output. Safety measures, such as proper grounding and insulation of electrical components, are critical to prevent any risk of electric shock or fire.

Beyond the physical setup, preparing your garden for Electroculture also involves a bit of ecological consideration. Understanding the local ecosystem and how your garden fits into it can help minimize any negative impacts while maximizing the benefits of Electroculture. For example, knowing which plants are native to your area and how they interact can help you choose crops that will thrive under Electroculture and contribute positively to the local biodiversity.

To further enhance the readiness of your garden for Electroculture, it is advisable to start small and scale up gradually. Begin by applying Electroculture techniques to a small section of your garden or a specific type of plant to monitor the effects and adjust the system as needed before expanding it to larger areas. This cautious approach allows for fine-tuning and can prevent widespread issues that might arise from a full-scale implementation without prior testing.

As you prepare your garden for Electroculture, keeping detailed records of all the changes made and the results observed can be incredibly valuable. Documenting soil conditions, plant responses, weather patterns, and any troubleshooting steps you take will provide a robust dataset that can inform future gardening efforts. This data can help refine Electroculture techniques, making them more effective and tailored to your specific garden conditions.

Preparing your garden for Electroculture is a meticulous process that requires assessing and enhancing soil conditions, carefully planning the garden layout and electrical setup, managing water levels, and considering the ecological impact. By methodically setting up and initially monitoring a small-scale Electroculture system, you can lay the foundation for the successful integration of this innovative technology into your garden, leading to healthier plants and more abundant yields.

Essential Tools and Materials

Setting up an Electroculture garden requires a range of specific tools and materials that facilitate the effective use of electrical currents to enhance plant growth. Understanding what these tools are and how they are used is crucial for anyone planning to implement this innovative agricultural technology. The selection of the right tools and materials ensures that the electrical system is not only effective but also safe and durable.

One of the primary materials needed for Electroculture is conductive wiring, typically made from copper due to its excellent electrical conductivity and resistance to corrosion. Copper wires are used to create a network of electrodes that can be buried in the soil or positioned around the plant beds. These wires are connected to a power source and arranged in specific patterns to optimize the distribution of the electric field across the garden. The size and gauge of the wire will depend on the garden's size and the specific requirements of the electrical system being installed.

In addition to copper wires, grounding rods are essential for any Electroculture setup. These rods are driven into the ground to help stabilize the electrical system and prevent any adverse effects from potential voltage spikes or electrical surges. Grounding rods are typically made of galvanized steel or copper and should be installed at strategic points within the garden to ensure the entire system is safely grounded.

A reliable power source is another critical component. For many gardeners, solar panels are an attractive option because they provide a sustainable and eco-friendly way to generate the electricity needed for Electroculture. Solar panels can be set up to charge a battery system, which in turn powers the Electroculture setup, ensuring that the garden receives a consistent and controlled flow of electricity. Alternatively, a low-voltage power supply from the main electrical grid can be used, but it must include a transformer or regulator to manage the output effectively.

To control and monitor the flow of electricity through the garden, you will need a series of switches, timers, and potentially a circuit breaker. These devices help manage when and how much electricity is sent through the wires, allowing for precise control over the electrical inputs that the plants receive. Timers are particularly useful for automating the system, ensuring that the plants are exposed to electrical stimulation at the most beneficial times, such as during specific growth phases.

Voltmeters and ammeters are crucial tools for monitoring the system's electrical parameters. Regular checks with these instruments help ensure that the system operates within the safe and effective range for plant stimulation. This monitoring is critical to prevent any damage to the plants or the electrical system from excessive voltage or amperage.

Waterproofing materials such as insulation tape and wire protectors are necessary to safeguard the electrical components from moisture and weather-related damage. These materials are used to insulate connections and protect exposed wires, maintaining the integrity of the system against environmental factors that could lead to shorts or corrosion.

Finally, basic gardening tools remain essential. Shovels, trowels, and gardening gloves are needed for the physical setup of the system, including digging trenches for the wires, planting, and routine garden maintenance. Having these tools at hand will make the installation process smoother and ensure that the garden is well-maintained, which is crucial for the success of any Electroculture project.

Gathering and effectively utilizing these tools and materials will equip you with the necessary resources to successfully set up and maintain an Electroculture garden. Each component plays a crucial role in ensuring that the electrical enhancements effectively stimulate plant growth while maintaining safety and sustainability. Properly assembled, these tools and materials pave the way for a highly productive and innovative agricultural practice that can revolutionize gardening outcomes.

Choosing the Right Copper Wires

Choosing the right type of copper wire is a crucial decision when setting up an Electroculture system, as it directly influences the effectiveness and efficiency of your garden's electrical enhancement. Copper wires come in various types and sizes, each suited to different garden setups and climate conditions. Understanding the specific characteristics and requirements of your garden will help you select the most appropriate copper wire, ensuring optimal electrical conductivity and durability.

The first consideration when choosing copper wires is the gauge, which refers to the thickness of the wire. Thicker wires (which have a lower gauge number) can carry more current and are less likely to heat up, making them ideal for larger gardens or systems that require longer wire lengths. For smaller gardens or lower power applications, thinner wires (higher gauge number) may be sufficient and can be more cost-effective. Typically, a gauge range from 12 to 18 is used in most garden applications, with 12 being on the thicker side and more robust for extensive or high-power setups.

The second factor to consider is the type of wire insulation. While bare copper wire is commonly used due to its excellent conductivity, it may not be the best choice for all environments. In areas with high humidity, frequent rainfall, or corrosive soil conditions, insulated copper wire is advisable. Insulation protects the copper from the elements, preventing corrosion and ensuring the longevity of the wire. The type of insulation can vary, but commonly used materials include PVC, rubber, or weather-resistant plastic. These materials provide a protective layer that withstands environmental stressors while maintaining the wire's performance.

Another important aspect is the wire's configuration. Solid copper wire, consisting of a single piece of copper, is generally more durable and resistant to breakage under stress, such as when buried in soil. However, stranded copper wire, which is made up of multiple small strands twisted together, offers more flexibility. This makes it easier to install in gardens with complex layouts or where wires need to be run around multiple obstacles. Stranded wire is also preferable in areas with seismic activity, as it can withstand vibrations and ground movements better than solid wire.

For gardeners in very cold climates, considerations about the wire's performance in low temperatures are also crucial. In freezing conditions, metal can become brittle and more prone to snapping. For such environments, selecting high-quality, insulated copper wire with a protective sheath that remains flexible in cold weather is essential. This ensures that the wire does not crack or break when the temperature drops, maintaining the integrity of your Electroculture system throughout the winter.

Additionally, the total length of the wire needed and its cost-effectiveness should be considered. Electroculture setups can require significant lengths of wire, especially for larger gardens. Calculating the total length needed for your specific garden layout and then comparing the costs of different wire types and gauges can help you make a budget-conscious choice without compromising on quality.

Selecting the right type of copper wire for your Electroculture system involves considering several factors: the wire's gauge, insulation type, configuration (solid or stranded), adaptability to climate conditions, and overall cost. By carefully evaluating each of these elements in relation to your garden's specific needs and environmental conditions, you can choose a copper wire that optimizes the effectiveness of your Electroculture setup, ensuring enhanced plant growth and productivity for years to come.

Soil Preparation and Testing

Preparing and testing the soil is a foundational step in setting up an Electroculture system. Proper soil preparation ensures that the electrical inputs utilized in Electroculture are effective, enhancing plant growth by improving nutrient uptake, moisture retention, and root development. Soil testing, on the other hand, provides critical information about the soil's current state, including its nutrient content, pH level, and electrical conductivity, all of which influence the effectiveness of Electroculture techniques.

Soil Preparation

The preparation of soil for Electroculture involves several steps aimed at optimizing its structure and composition. First, it's crucial to ensure that the soil has adequate drainage while still retaining enough moisture to conduct electricity effectively. This can be achieved by adjusting the soil's texture and structure. For soils that are too dense or clayey, incorporating organic matter such as compost or peat can improve aeration and drainage. In contrast, sandy soils, which drain too quickly, may benefit from the addition of clay minerals or organic matter to enhance their moisture-holding capacity.

Secondly, the incorporation of amendments based on the initial soil test results can adjust nutrient levels and pH to create an ideal growing environment. If the soil is too acidic or alkaline, adding lime or sulfur, respectively, can bring the pH to a level that maximizes nutrient availability and root development. Similarly, if the soil tests show deficiencies in key nutrients like nitrogen, phosphorus, or potassium, appropriate fertilizers should be added to replenish these essential elements.

Soil Testing

Testing the soil is a critical step that should precede the actual installation of an Electroculture system. This testing can be done through DIY soil test kits available at garden centers or by sending soil samples to a laboratory for more detailed analysis. The primary aspects of testing include the following:

- **pH Level:** The soil's pH affects nutrient availability. Most plants prefer a slightly acidic to neutral pH (between 6.0 and 7.0). Knowing the pH can help in adjusting it to suit the specific needs of the plants being grown.
- **Nutrient Profile:** Testing for key nutrients—nitrogen, phosphorus, potassium (NPK), and trace elements like iron and magnesium—provides insights into what fertilizers or amendments are needed.
- **Electrical Conductivity (EC):** This test measures the soil's ability to conduct electrical currents, which is crucial for Electroculture. A higher EC indicates that the soil can effectively conduct electrical charges, aiding in the stimulation of plant growth. If EC levels are low, it may be necessary to adjust soil components or moisture levels to improve conductivity.
- **Organic Matter Content:** High organic matter improves soil structure, nutrient retention, and microbial activity, all beneficial for Electroculture.

- **Moisture Content:** Consistent moisture levels are necessary for effective Electroculture, as water is a conductor of electricity. Testing soil moisture helps in setting up irrigation systems that maintain optimal moisture levels without over-saturating the soil.

After conducting these tests, the results should guide the preparation adjustments. For instance, if the soil conductivity is lower than desired, adding natural salts or moisture-retentive amendments can help. If nutrient levels are off-balance, specific fertilizers can be incorporated to meet the plant's needs.

Continuous Monitoring and Adjustment

Once the Electroculture system is operational, continuous monitoring and periodic re-testing of the soil are advisable. This ongoing process helps in adjusting the system based on changing soil conditions and plant requirements. It ensures that the soil remains conducive to Electroculture, thereby maximizing the growth potential and health of the plants.

Careful soil preparation and thorough testing form the bedrock of a successful Electroculture setup. By understanding and optimizing soil conditions, gardeners and farmers can create an environment that leverages electrical enhancements for superior plant growth. This process not only boosts yields but also contributes to the sustainability of agricultural practices by ensuring that resources are used efficiently and responsibly.

Understanding Soil Types

Understanding how different soil types affect Electroculture is crucial for effectively utilizing this technology in agriculture. Soil type can influence the movement of electricity through the soil, the availability of nutrients, and the overall health of the plants. By comprehending the characteristics of various soil types, gardeners and farmers can better tailor their Electroculture strategies to enhance plant growth and productivity.

Characteristics of Major Soil Types

Sandy Soil: Sandy soil is characterized by large, coarse particles that provide excellent drainage and aeration. However, its ability to hold nutrients and water is relatively poor. In the context of Electroculture, sandy soil can pose a challenge due to its low electrical conductivity. The large air spaces between the sand particles make it difficult for electrical currents to pass through effectively. To improve its suitability for Electroculture, organic matter, clay, or loam can be added to increase water retention and conductivity.

Clay Soil: Clay soil consists of very fine particles that are closely packed together, leading to high water retention but poor drainage and aeration. Its dense nature allows it to conduct electricity better than sandy soil, which can be advantageous for Electroculture. However, poor aeration and drainage can be detrimental to root health. To mitigate these issues, incorporating organic matter or sand can improve the structure, enhance drainage, and decrease density to prevent overly rapid electrical transmission that might stress plant roots.

Loamy Soil: Loamy soil is considered ideal for most forms of agriculture, including Electroculture because it balances the properties of sand, silt, and clay. It retains nutrients and moisture well while still providing good drainage and aeration. The balanced texture of loamy soil supports moderate electrical conductivity, which is beneficial for Electroculture as it ensures even distribution of electric fields without over-concentration that could harm plant roots.

Peaty Soil: Peaty soil is rich in organic matter and typically has a high water content, which can enhance its electrical conductivity. This type of soil is good for Electroculture, especially in terms of nutrient supply and moisture retention. However, peaty soil tends to be more acidic, which might limit the availability of certain nutrients. Adjusting the pH with lime or other amendments can make this soil type more conducive to Electroculture.

Silty Soil: Silty soil, like loamy soil, has good moisture retention and sufficient aeration. It is composed of finer particles than sand but is not as compact as clay, offering a balance that is generally favorable for plant growth and Electroculture. Silty soil's moderate electrical conductivity and ability to retain nutrients make it suitable for Electroculture, though it may require regular monitoring to ensure that the moisture levels do not become excessive.

Adjusting Soil Types for Electroculture

For effective Electroculture, it is often necessary to adjust the natural soil type to optimize its properties. This may involve:

- **Enhancing Electrical Conductivity:** For sandy and some silty soils, increasing organic content or adding clay can improve conductivity. For clay soils, adjustments might focus more on improving structure and reducing density to control the flow of electricity.
- **Balancing pH Levels:** Since the availability of nutrients is pH-dependent and Electroculture can sometimes affect soil pH, it is important to monitor and adjust pH levels to maintain optimal conditions for nutrient uptake.
- **Improving Nutrient Availability:** Incorporating compost, manure, or commercial fertilizers can help adjust the nutrient profile of the soil to meet specific plant needs, which is particularly crucial in soils like sandy or peaty types that may inherently lack certain nutrients.
- **Regulating Moisture Content:** Installing proper drainage systems in clay and peaty soils or using irrigation systems like drip or sub-irrigation can help maintain optimal moisture levels, ensuring that electrical currents are effectively conducted without saturating the soil.

By understanding the interaction between different soil types and Electroculture, gardeners, and farmers can make informed decisions on how to prepare and manage their soil. This knowledge allows for the customization of Electroculture practices to suit specific environmental conditions and crop requirements, leading to more successful agricultural outcomes.

Testing Soil pH and Nutrients

Testing soil pH and nutrients is crucial for any gardener or farmer, especially when preparing for the implementation of Electroculture, as it directly affects plant health and the effectiveness of electrical stimulation. Here's how to conduct these tests and adjust the soil accordingly to optimize your garden's environment for plant growth.

Begin by collecting soil samples from various locations around your garden or farm to get a comprehensive overview of the soil conditions. It's important to collect samples at a consistent depth, typically 6 to 8 inches below the surface for gardens and deeper for larger agricultural fields. Use a clean tool to avoid contamination of the samples and mix soil from several spots to create a composite sample for testing.

To test soil pH, you can use a soil pH meter by inserting the probe into the soil sample. Ensure the soil is moist, as this will give a more accurate reading. Follow the manufacturer's instructions for your specific pH meter. Alternatively, pH testing kits are available, which typically include a test tube, testing solution, and a color chart. Place a small sample of your soil into the test tube, add the provided solution, and shake it well determine the pH level by comparing the color of the solution to the color chart after a few minutes. You can submit samples of your soil to a lab for a more thorough study. In-depth details regarding pH values, nitrogen content, and other soil properties will be provided by this.

Once you have your pH results, adjust the soil pH accordingly. To raise the pH of the soil if it is excessively acidic (pH less than 6), apply lime (calcium carbonate). The kind of soil and pH level will determine how much lime is needed; sandy soils may require less lime, while clay soils may require more. In order to lower the pH of the soil if it is too alkaline (pH greater than 7.5), add sulfur or aluminum sulfate. The amount needed will depend on the soil type and how much the pH needs to be adjusted.

For nutrient testing, home testing kits can measure levels of key nutrients such as nitrogen (N), phosphorus (P), and potassium (K). These kits usually involve adding soil to a chemical solution and comparing the color change to a chart. when assessing soil nutrients, sending a sample to a soil testing facility produces more precise findings. In addition to NPK, this can provide a comprehensive analysis of secondary nutrients and micronutrients, which are equally vital to plant health.

Based on the nutrient test results, adjust nutrient levels as needed. For nitrogen deficiency, add organic matter such as compost, manure, or a suitable nitrogen fertilizer. Slow-release fertilizers are preferred as they do not wash away easily, providing a steady supply of nitrogen. If the soil is low in phosphorus, add rock phosphate or bone meal, which are effective sources of phosphorus for plants. For potassium deficiency, add potash or a potassium sulfate fertilizer to increase potassium levels, which is especially important for plants that are heavy feeders of potassium.

Regular monitoring of soil pH and nutrient levels is vital, as soil conditions can change due to factors such as rainfall, crop uptake, and the addition of amendments. Annual or biannual testing is recommended to keep track of soil health and make adjustments as needed. This systematic approach not only prepares the soil for successful Electroculture but also ensures ongoing soil health and fertility, which is crucial for sustainable agricultural practices. By maintaining the right balance of pH and nutrients, you optimize conditions for electrical conductivity and plant growth, leading to a more productive and efficient garden or farm.

Amending Soil for Optimal Growth

Amending soil to support Electroculture involves a series of deliberate steps aimed at enhancing soil properties to optimize plant growth and electrical conductivity. This process is crucial because the efficacy of Electroculture largely depends on the soil's ability to conduct electricity effectively and support robust plant health. Here are some techniques for amending soil to create an ideal environment for Electroculture.

Improving the physical structure of the soil is fundamental to promoting good root development and efficient electrical conduction. For soils that are too compacted, like clay soils, incorporating organic matter such as compost, peat moss, or aged manure can help to loosen and aerate the soil. This not only improves drainage and aeration but also enhances the soil's capacity to hold nutrients and water, which are essential for plant growth.

For sandy soils, which typically have large particles and poor water-holding capacity, adding organic matter or clay particles can help to increase moisture retention and improve the soil's structure. This adjustment makes the soil more conducive to Electroculture by maintaining moisture levels that are necessary for effective electrical conduction.

Soil pH can significantly affect the mobility of ions in the soil, which in turn influences the soil's electrical properties and nutrient availability. Most plants prefer a slightly acidic to neutral pH (between 6.0 and 7.0). For soils that are too acidic, adding lime (calcium carbonate) can help to raise the pH, while sulfur or aluminum sulfate can be used to lower the pH of alkaline soils. Adjusting the pH to the optimal range for the specific plants being grown will enhance both the effectiveness of Electroculture and the overall soil health.

Since Electroculture relies on the movement of electrical currents through the soil, enhancing the soil's electrical conductivity is crucial. This can be achieved by maintaining adequate moisture levels and adding conductive amendments. For instance, adding small amounts of natural salts (like Epsom salts or rock dust) can increase ion content in the soil, improving conductivity. However, it is essential to use salts sparingly to avoid damaging plants and degrading soil health.

Another method to improve conductivity is by incorporating biochar into the soil. Biochar is a form of charcoal used to improve soil fertility and increase its carbon content. It not only enhances soil structure and nutrient retention but also increases soil conductivity due to its porous nature and high surface area.

Electroculture can stimulate rapid plant growth, which may increase nutrient demands. Ensuring the soil is rich in essential nutrients is therefore critical. Conduct a soil test to identify any nutrient deficiencies and address them with appropriate organic or inorganic fertilizers. Focus particularly on macronutrients like nitrogen, phosphorus, and potassium, as well as important micronutrients such as magnesium and calcium, which play critical roles in plant health and growth.

The presence of beneficial microbes in the soil can significantly enhance plant growth and soil health. These microbes aid in nutrient breakdown and absorption, enhance soil structure, and help suppress soil-borne diseases. Incorporating compost or compost tea, planting cover crops, and applying microbial inoculants are effective ways to increase microbial diversity and activity in the soil.

After implementing these amendments, it is crucial to monitor soil conditions regularly. This includes checking pH levels, moisture content, nutrient status, and overall soil structure. Regular monitoring allows for timely adjustments to maintain optimal conditions for Electroculture.

By thoroughly amending the soil as described, you create an environment that supports not only Electroculture but also the broader health and productivity of your garden or farm. This holistic approach to soil management ensures that the benefits of Electroculture are fully realized, leading to healthier plants and more abundant yields.

3

INSTALLING COPPER ANTENNA WIRES

Selecting the Perfect Antenna Design

Selecting the perfect antenna design for an Electroculture system is crucial for maximizing the effectiveness of electrical inputs across different types of gardens. The design of the antenna influences how evenly and effectively the electric field is distributed in the soil, impacting plant growth, nutrient uptake, and overall health. Understanding the various antenna designs and their specific benefits can help gardeners and farmers choose the right configuration to suit their unique agricultural needs.

The basic principle behind using antennas in Electroculture involves the installation of conductive materials, typically copper wires, into the garden. These antennas can take various forms depending on the size of the garden, the type of crops, and specific environmental conditions. The goal is to create a network that delivers electric currents uniformly, stimulating plant growth without causing any harmful effects.

One common antenna design is the grid layout, where wires are laid out in a square or rectangular grid pattern beneath the soil surface. This design is particularly effective for large, open areas where uniformity in electric field distribution is crucial. The grid layout ensures that all plants receive an equal amount of stimulation, promoting consistent growth across the entire garden. This design is versatile and can be scaled up or down, making it suitable for both commercial farms and smaller residential gardens.

Another popular design is the radial layout, where wires radiate outward from a central point. This design is beneficial for gardens where plants are arranged in circular or semi-circular patterns, such as in orchards or rounded garden beds. The radial layout helps focus the electric field more intensely near the center, gradually decreasing towards the edges. This is particularly useful for boosting the growth of plants located at the center of the garden, which might otherwise receive less attention and resources.

For row crops, a parallel layout can be more appropriate. In this design, wires are run parallel to the rows of plants. This ensures that the electric fields are aligned with the row direction, providing a more targeted and efficient stimulation of the plant's root systems. The parallel layout is ideal for long, narrow gardens or fields where crops are planted in rows, such as in vineyards or vegetable farms.

Some advanced Electroculture systems utilize a spiral or helical antenna design, which can be particularly effective for dynamic environments or specific types of crops. The spiral design generates a concentrated electric field that can be beneficial for stimulating deep-rooted plants. It also allows for a more concentrated delivery of electric energy to areas that require more intense stimulation, such as the root zones of particularly high-value crops.

The zigzag layout is another innovative design, especially useful in terraced gardens or sloped areas. In this design, the wires are laid out in a zigzag pattern across the slope, helping to evenly distribute the electric field across a gradient. This prevents runoff of electrically enhanced water and ensures that all parts of the slope are equally stimulated, which is important for preventing erosion and promoting uniform plant growth.

In addition to these designs, custom layouts can be created to address specific challenges or goals. For example, areas with irregular shapes or mixed planting styles might require a combination of grid and radial layouts to effectively cover all areas. The flexibility to adapt or combine different designs allows Electroculture practitioners to tailor their systems to the precise needs of their gardens or farms.

When selecting the antenna design, it is also important to consider the type of soil, as its conductivity can affect how well the electric fields are transmitted. Soils with higher clay content may require different wire spacing or depth compared to sandy soils. Furthermore, environmental factors such as climate, moisture levels, and the presence of natural electromagnetic fields should also influence the design choice.

Ultimately, selecting the perfect antenna design is a critical step in setting up an effective Electroculture system. By carefully considering the type of garden, crop arrangement, soil characteristics, and environmental conditions, gardeners and farmers can optimize their use of electrical stimulation to enhance plant growth. This careful planning ensures that Electroculture becomes a valuable tool in achieving higher yields and healthier plants in a sustainable and environmentally friendly manner.

Step-by-Step Installation Guide

Installing an antenna system for Electroculture involves a series of detailed steps that ensure the proper setup and functionality of the electrical enhancements in your garden or farm. This step-by-step installation guide will walk you through the process from the initial preparation to the final setup, providing practical advice to make the installation as smooth and effective as possible.

Before laying down any wires or connecting any electrical components, it is crucial to prepare the site. This involves clearing the area of large debris, rocks, and weeds that might interfere with the installation. You should also mark out the exact locations where the antennas will be laid according to the chosen design pattern (grid, radial, parallel, etc.). Use flags or spray paint for clear, visible marking.

Once the area is prepared, measure the lengths of copper wire required based on your predetermined layout. Ensure you have a little extra wire to accommodate any adjustments needed during installation. Cut the copper wire using appropriate tools—wire cutters for smaller gauge wires or a heavy-duty cable cutter for larger gauges.

Begin laying the wires following the marks you have set out. If using a grid or parallel layout, start from one end of the area and roll out the wire in straight lines, securing it with U-shaped garden staples every few feet to keep it in place. For radial or spiral designs, start from the central point and work your way outward, securing the wire as you go. Ensure that the wires are laid flat and evenly against the soil to avoid any irregularities that could affect the distribution of the electric field.

Once all wires are laid out and secured, they need to be buried. The depth at which you bury the wires depends on the type of soil and the root depth of the plants you are growing. Generally, a depth of 4-6 inches is recommended. Use a garden spade or a powered trencher for larger areas to dig a shallow trench along the lines where the wires are laid. Carefully place the wires in the trench and cover them with soil, packing it down lightly to remove air pockets.

With the wires in place, the next step is to connect them to a power source. This typically involves running a main feeder wire from a control box (which houses the power source and any timers or regulators) to the start of your antenna system. Make sure all electrical connections are made using waterproof connectors and that all exposed wiring is covered with conduit or buried to protect it from the elements and accidental damage.

For safety and system effectiveness, installing a grounding rod is essential. Drive a copper grounding rod at least 6 feet into the ground near the control box. Connect the grounding rod to the system using heavy gauge grounding wire and clamps to ensure a secure and stable electrical grounding.

Before using the system, it's important to test it to ensure everything is functioning correctly. Connect the power supply and use a voltmeter and ammeter to test the voltage and current along different points of the antenna system. This will help you verify that the electrical current is evenly distributed and within safe operational limits. Make any necessary adjustments to the layout or connections based on these readings.

After the initial testing, monitor the system over the first few days to make adjustments as necessary. Check for any signs of plant stress or irregular growth, which could indicate issues with the electric field strength or coverage. Adjust the power settings or the wire layout as needed to optimize the system's performance.

Regular maintenance of the Electroculture system is vital to ensure its longevity and effectiveness. Periodically check the condition of the wires, connectors, and the power supply unit for any signs of wear, corrosion, or damage. Also, re-test the system with a voltmeter and ammeter at the start of each growing season to ensure that it remains in good working order.

By following these detailed steps, you can successfully install an antenna system for Electroculture, providing your plants with an optimal growing environment enhanced by electrical stimulation. This careful setup ensures that your garden or farm can reap the maximum benefits from this innovative agricultural technology.

Positioning Antennas for Maximum Effectiveness

Positioning antennas effectively in an Electroculture system is critical to maximizing the benefits of electrical stimulation for plant growth. The strategic placement of these antennas ensures that the electric fields are distributed uniformly and efficiently, providing optimal stimulation to the root systems of all plants within the area. Here are some key strategies to consider when positioning antennas to achieve the best results in your garden or farm.

Before positioning antennas, it's essential to have a detailed understanding of your garden's layout. This includes knowing the size and shape of the planting area, the types of crops planted, and their specific growth patterns and root systems. Different plants may have varying requirements and sensitivities to electrical fields, so understanding these nuances can help tailor the antenna placement to suit specific needs.

Soil type plays a significant role in how well electrical currents are conducted. Heavier clay soils, which have better moisture retention and electrical conductivity, may require less dense antenna placement compared to sandy soils, which are poorer conductors. Knowing your soil type and its properties can help you adjust the spacing and depth of your antennas to ensure effective electrical distribution.

The depth at which the antennas (wires) are buried affects how the electrical currents influence plant roots. Generally, wires should be placed close enough to the root zone to affect the roots but not so close that they could cause damage. For most crops, burying wires about 4-6 inches below the surface is ideal. However, for larger plants with deeper root systems, you may need to place the wires deeper into the soil.

For large, uniform areas, using a grid pattern is often the most effective way to position antennas. This pattern ensures that the electrical fields are evenly distributed across the entire area, providing consistent stimulation to all plants. In a grid layout, wires are laid out in a matrix of vertical and horizontal lines that intersect, creating a series of squares or rectangles that cover the whole garden bed.

In gardens where plants are arranged in circular or semi-circular patterns, such as in orchards or around central features, a radial antenna pattern might be more effective. In this setup, wires radiate out from a central point, like spokes on a wheel, which helps focus the electric field more intensely towards the center and diminishes outward, matching the natural layout of the plants.

Different crops might require different levels of exposure to electric fields. For example, fast-growing vegetables might benefit from more intense electrical stimulation, while perennial plants might require a gentler approach. Adjusting the spacing between the wires can help modulate the intensity of the electric fields experienced by different types of plants, optimizing growth according to each plant's specific requirements.

Regularly monitor the effects of the Electroculture system on your plants and adjust the positioning of your antennas based on this feedback. If certain areas of your garden are showing less vigorous growth or signs of stress, it might be necessary to reposition the wires closer to or further from these areas. This kind of responsive management can significantly enhance the overall effectiveness of your Electroculture setup.

Modern technology, such as GPS mapping and soil conductivity meters, can be invaluable in planning and implementing precise antenna layouts. These tools can help you visualize the electric field distribution and make adjustments in real time, ensuring that every part of your garden receives optimal stimulation.

By carefully considering these strategies when positioning antennas, gardeners and farmers can significantly enhance the effectiveness of their Electroculture systems. Proper antenna placement tailored to the specific conditions and requirements of your garden ensures that all plants benefit from enhanced growth, increased yields, and improved overall health.

Integrating Antennas with Existing Structures

Integrating Electroculture systems into existing garden setups can be a nuanced task, especially when aiming to enhance growth without disrupting the established ecosystem. Electroculture, which utilizes electrical stimulation to promote plant health and productivity, must be incorporated thoughtfully to ensure it complements existing structures and planting patterns. Here are some insights and techniques for effectively blending Electroculture systems with your current garden setup.

The first step in integrating an Electroculture system is to thoroughly understand the current layout and specific needs of your garden. This includes knowing the types of plants you have, their growth stages, root systems, and the arrangement of any physical structures like trellises, paths, or irrigation systems. A detailed map or sketch of the garden can be invaluable here, providing a visual reference as you plan the installation of your Electroculture system.

Since Electroculture relies heavily on the conductivity of the soil to be effective, assessing the existing soil conditions is crucial. Testing for moisture content, pH levels, and nutrient density can help determine how to best integrate the electrical components without harming the garden's current ecosystem. This might involve amending the soil to improve its conductivity or adjusting the depth at which the wires are installed based on the specific root depths of your plants.

Selecting the right antenna design is key to a seamless integration. For gardens with extensive raised beds or container plants, consider using individual wire setups tailored to each section rather than a single large grid. This approach reduces interference with existing root systems and allows for targeted stimulation, which is especially beneficial for gardens with a diverse range of plant species and sizes.

In landscapes where plants are organized in rows, such as in vegetable gardens or vineyards, parallel wire systems can be ideal. These can be installed along the rows, mirroring the natural layout and ensuring that the electrical stimulation directly benefits the root zones of the plants without unnecessary disruption to the soil structure.

When installing Electroculture wires in an established garden, use minimally invasive techniques to avoid significant disruption to the plants. Trenching tools can be used to insert wires below the surface with minimal disturbance to the roots. For raised beds or densely planted areas, flexible drill bits or soil probes might be used to create narrow channels for the wires, preserving the integrity of the planting medium.

Many gardens already have irrigation systems in place, which must be considered when integrating Electroculture. The irrigation scheme and the Electroculture wires should work together harmoniously so that the two systems can function independently of one another. In some cases, it may be beneficial to combine the irrigation and Electroculture systems into a single infrastructure, where water and electrical stimulation are delivered simultaneously, enhancing the efficiency of both.

When incorporating an Electroculture system, ensuring electrical safety is paramount. This includes using insulated wires, proper grounding techniques, and waterproof connections to prevent any risk of short circuits or electrical leakage, particularly in wet environments. Additionally, all electrical connections should be housed in protective enclosures and clearly marked to avoid accidental damage during garden maintenance activities.

For gardens that are not only functional but also designed for aesthetic enjoyment, it's important that the Electroculture system does not detract from the visual appeal. Wires should be buried or discreetly camouflaged among plants. For ornamental gardens, consider using decorative covers or integrating the wiring along discrete pathways or behind structures to keep them out of direct sight.

Once the Electroculture system is integrated, continuous monitoring and adjustments are essential to ensure it works harmoniously with the existing garden setup. This involves regular checks of electrical outputs, observing plant responses, and making adjustments to wire placement or power settings as needed. Feedback from these observations will guide any further integration efforts, ensuring the Electroculture system remains a beneficial addition to the garden.

Integrating an Electroculture system into an existing garden setup requires careful planning and consideration of both the technical and practical aspects of garden management. Gardeners can use Electroculture to increase plant growth and productivity while preserving the organic balance and aesthetic appeal of their landscapes by employing a deliberate and well-informed strategy.

Safety Considerations

Implementing an Electroculture system, which uses electrical currents to stimulate plant growth, requires careful attention to safety considerations to prevent accidents and ensure the system operates effectively and safely. This comprehensive overview covers essential safety tips and practices to safeguard gardeners, the environment, and the infrastructure of the Electroculture setup itself.

Understanding Electrical Safety Basics

The foundational aspect of safely operating an Electroculture system is understanding the basics of electrical safety. It's crucial to be aware that even low-voltage systems can pose hazards if not properly managed. Basic knowledge about electrical current, grounding, insulation, and safe handling practices is essential. Those installing or maintaining the system should either have this knowledge or consult with a professional electrician. It's also advisable to thoroughly read and follow the manufacturer's instructions when setting up and operating Electroculture equipment.

Proper Installation of Electrical Components

Proper installation of electrical components is key to preventing malfunctions and accidents. All wiring should comply with local electrical codes and standards, which are designed to minimize the risk of electrical fires and other hazards. Use only wires and components that are rated for outdoor use and appropriate for the electrical load they will carry. Ensure that all connections are tight and secure to prevent arcing, which can lead to fires.

Use of Ground Fault Circuit Interrupters (GFCIs)

To enhance safety, especially in outdoor environments where the presence of water increases the risk of electrical shock, installing Ground Fault Circuit Interrupters (GFCIs) is crucial. GFCIs can detect imbalances in the electrical current and shut down the power supply in milliseconds, significantly reducing the risk of shock. This is particularly important for Electroculture systems, where irrigation and dew can create wet conditions conducive to electrical hazards.

Regular Inspection and Maintenance

Regular inspections are vital to ensuring the ongoing safety and effectiveness of an Electroculture system. Check for signs of wear and tear, such as frayed wires, corroded connectors, or damaged insulation. These issues should be addressed immediately to prevent more serious problems. It's also important to ensure that all electrical components remain securely mounted and protected from environmental elements like wind, rain, and soil erosion.

Proper Grounding Techniques

Effective grounding is one of the most critical safety measures for any electrical system. In Electroculture, grounding not only protects against electrical shock but also enhances the effectiveness of the system. Ensure that grounding rods are installed according to the guidelines and are connected with appropriate grounding wires. This setup should be checked periodically to ensure that it remains functional and secure.

Safe Handling and Operation

When working with or near the Electroculture system, always wear rubber-soled shoes and use insulated tools to minimize the risk of electrical shock. Avoid handling electrical components with wet hands or standing in water, and never attempt to make adjustments to the system during wet weather conditions. Additionally, ensure that the system is turned off when performing maintenance or adjustments.

Training and Awareness

Anyone involved in installing or maintaining an Electroculture system should have basic training in electrical safety. This includes understanding how to safely handle electrical components, recognizing the signs of electrical problems, and knowing how to respond in an emergency. Awareness campaigns or regular safety briefings can be beneficial for groups or teams working on larger installations.

Protecting Plants and Wildlife

While focusing on human safety, it's also important to consider the safety of the plants and surrounding wildlife. Ensure that the electrical outputs used are within safe limits for plant health, as excessive electrical stimulation can damage plants. Additionally, it is important to position wires and equipment where they will not pose hazards to animals—both wildlife and domestic pets.

Emergency Preparedness

Have a clear plan in place for dealing with electrical emergencies, which should include knowing how to quickly cut power to the system, administer first aid for electrical shock, and contact emergency services. Regularly review and practice the emergency plan with all individuals involved in managing the Electroculture system.

By adhering to these safety considerations, gardeners and farmers can safely enjoy the benefits of Electroculture, which improves plant growth and productivity without compromising the safety of people, plants, or the environment. Maintaining a diligent focus on safety not only prevents accidents but also ensures that the system remains functional and effective over the long term, providing sustained benefits in agricultural settings.

Are You Enjoying the Journey So Far?

As you delve deeper into the world of electrophilic gardening, we hope the techniques and insights are proving useful. Your feedback is incredibly important to us and to the community of readers who are considering this innovative approach to gardening. Please consider leaving a review on Amazon to let us and others know how the book is helping you grow!

How You Can Share Your Review:

Through Amazon.com:

- Go to the Amazon page where you found my book.
- Navigate to the 'Customer Reviews' section.
- Click on 'Write a customer review' to share your valuable insights.
- Instant QR Code Access: Simply scan the QR code below with your smartphone to be directed to the Amazon review section.

4

TROUBLESHOOTING COMMON ISSUES

Diagnosing and Addressing Electroculture Challenges

Electroculture offers significant benefits for plant growth and productivity, but like any technology, it may encounter challenges that could hinder its effectiveness. Understanding common issues and knowing how to resolve them is crucial for maintaining a thriving Electroculture system. This discussion explores various issues that might arise and provides practical solutions to ensure your garden continues to thrive under Electroculture.

One of the most common challenges in Electroculture is electrical system failures, which can stem from faulty wiring, inadequate power supply, or improper connections. These failures not only interrupt the stimulation of plants but can also pose safety risks. To address this, regularly inspect the electrical components of your system for any signs of wear, corrosion, or damage. Make sure all connections are secure and that wires are properly insulated and protected from environmental elements. Using a multimeter, regularly check the voltage and current to ensure they are within the desired range. If any irregularities are found, shut down the system and address the issue immediately, replacing faulty components as necessary.

Another issue that may arise is inconsistent growth among plants, which could be due to an uneven distribution of the electric field. This might result from incorrect wire placement, varying soil conditions, or differing moisture levels across the garden. To solve this, ensure that the antenna layout evenly distributes the electrical current across your garden. This might involve adjusting the layout or adding more wires to areas where growth seems stunted. Also, conducting regular soil tests can help identify and rectify areas with poor conductivity or nutrient imbalances. Adjusting soil moisture levels to maintain consistent dampness can also help stabilize the effectiveness of the electric field.

Overstimulation of plants can occur when they are exposed to too intense or too frequent electrical stimulation, which can stress the plants, leading to yellowing leaves, stunted growth, or even plant death. To manage this, monitor the health of your plants regularly and adjust the voltage or frequency of the electrical input accordingly. It's crucial to begin with lower settings and gradually increase the intensity while monitoring the plants' reactions. Using timers to control the duration and frequency of stimulation can help manage this issue effectively.

The metal components used in Electroculture, particularly the copper wires, can corrode over time, especially in acidic soils or in environments with high humidity, which can impair the system's functionality. To prevent this, choose high-quality, corrosion-resistant materials when setting up your system. Regularly check for corrosion and replace any damaged parts immediately. Applying corrosion inhibitors or using wires with corrosion-resistant coatings can also prolong the lifespan of your system components.

Electrical leaks and safety hazards can occur if the insulation on wires deteriorates or if connections are not waterproofed. Regular inspections can help catch and repair insulation damage or faulty connections early. Ensure that all outdoor electrical connections are properly sealed against moisture. Installing Ground Fault Circuit Interrupters (GFCIs), which instantly cut the current when a leak is detected, can provide an additional layer of security.

Electroculture might be viewed with skepticism by neighbors or community members who are unfamiliar with the technique and concerned about the use of electricity in gardens. To address this, educate those around you about the benefits and safety features of Electroculture. Demonstrating the system's safety precautions and explaining how it improves plant growth and reduces chemical use can help alleviate concerns and foster community acceptance.

As Electroculture systems scale from small gardens to larger agricultural operations, issues such as power distribution, system management, and maintenance can become more complex. For larger systems, consider consulting with or hiring a professional with experience in Electroculture and agricultural electrical systems. This can ensure that the scale-up is managed efficiently and that the system remains reliable and effective.

By addressing these common Electroculture challenges with informed and thoughtful solutions, gardeners and farmers can enhance the resilience and productivity of their Electroculture systems. Regular monitoring, maintenance, and adjustments based on ongoing observations and testing are key to overcoming obstacles and achieving sustained success in Electroculture gardening.

Fine-Tuning Your Setup for Maximum Efficiency

Fine-tuning an Electroculture system is essential for maximizing its efficiency and ensuring the best possible outcomes for plant growth. Making thoughtful adjustments to the system can help optimize its performance, enhance plant health, and increase yield without compromising safety. Here are several tips for refining your Electroculture setup to achieve maximum efficiency.

First, keep a careful eye on the electrical parameters. You should modify the voltage and current in accordance with the particular requirements of your plants and the state of the soil. Start with lower settings to avoid stressing the plants and gradually increase the intensity if the plants respond well. Use a multimeter regularly to check that the electrical outputs are within safe and effective ranges. This ensures that the plants receive the right amount of stimulation without the risk of overexposure, which can be detrimental to their health.

Secondly, reassess the layout of your Electroculture wires. The initial setup might need adjustments as plants grow and their root systems expand. Make sure that the wires are optimally placed relative to the plant roots to enhance the effectiveness of the electrical stimulation. Sometimes, repositioning the wires slightly deeper or closer to specific root zones can significantly improve the system's impact.

Additionally, consider the timing and duration of electrical stimulation. Plants may benefit from varying the frequency and timing of the electric currents depending on their growth stage. For example, younger plants might need less frequent stimulation, while flowering plants might benefit from more frequent exposure. Implementing a timer can automate this process, making it easier to manage and more precise.

An Electroculture system's ability to function effectively is greatly influenced by the state of the soil. Check the pH, nutritional levels, and electrical conductivity of the soil on a regular basis. Regularly test the soil for electrical conductivity, pH, and nutrient levels. If the soil becomes too dry, its ability to conduct electricity effectively can decrease, reducing the efficiency of the system. Conversely, overly wet soil might lead to uneven distribution of the electric fields. Adjust your irrigation practices to maintain optimal soil moisture levels that support good conductivity without waterlogging the plants.

Furthermore, integrating feedback mechanisms into your system can provide valuable insights into its performance. Sensors that monitor soil moisture, temperature, and electrical conductivity can give you real-time data on the conditions in your garden. This information can be used to make immediate adjustments to the system, such as modifying the electrical input or changing irrigation schedules, ensuring that the plants always have optimal growing conditions.

Another aspect of fine-tuning involves addressing any interference from external sources. Electroculture systems can be sensitive to interference from large metal objects or other electronic devices. Ensure that your system is isolated from such interferences, possibly repositioning parts of the system if necessary. Shielding wires and components can also help minimize potential disruptions caused by external electromagnetic fields.

Regular maintenance is also key to keeping the Electroculture system running at maximum efficiency. Regularly inspect all system components for wear and tear indicators, such as loose connections, corrosion, or damaged wires. As soon as possible, replace or fix any broken components to keep the system functional and intact.

Lastly, it's critical to always be learning and adapting. Keep up with the most recent findings and developments in Electroculture technology. Try out novel methods or materials to see if they can enhance your system's functionality. Experiment with new techniques or materials that might improve the performance of your system. Sharing experiences with other Electroculture practitioners can also provide new insights and ideas for enhancing your setup.

By implementing these fine-tuning strategies, you can significantly enhance the performance of your Electroculture system. Your Electroculture setup will remain efficient, effective, and safe with regular monitoring and modifications based on your plants' unique demands and the conditions of your garden. This will result in healthier plants and more abundant outputs.

Sustainable Practices and Organic Solutions

Electroculture, which leverages electrical currents to promote plant growth, aligns well with sustainable agricultural practices and the principles of organic farming. By combining Electroculture with organic methods, gardeners and farmers can enhance plant productivity and health without reliance on chemical fertilizers and pesticides, thus minimizing environmental impact. Here's how to effectively incorporate sustainable practices and organic solutions within Electroculture setups.

The foundation of integrating sustainable methods in Electroculture starts with the source of electricity. Utilizing renewable energy sources, such as solar panels or wind turbines, to power Electroculture systems ensures that the operations are environmentally friendly and carbon-neutral. These renewable sources provide a consistent and cost-effective power supply, reducing dependency on non-renewable energy resources and enhancing the sustainability of agricultural practices.

Water conservation is an important part of sustainable Electroculture. Given that water conducts electricity, maintaining optimal soil moisture is crucial not just for plant health but also for the effectiveness of Electroculture. Drip irrigation systems can be integrated with Electroculture setups to regulate water usage precisely. These systems deliver water directly to the root zones, where it is most needed, reducing waste and eliminating overwatering, which can lead to nutrient loss and soil erosion.

Soil health is paramount in any organic farming operation, and this is no less true within Electroculture. Preserving soil integrity by avoiding the use of chemical fertilizers and pesticides is essential. Instead, organic amendments such as compost, manure, or biochar can be used to enrich the soil. These organic materials improve soil structure, enhance microbial activity, and gradually release nutrients, thereby supporting plant growth naturally. They also increase the soil's carbon content, which can improve its capacity to conduct electrical currents effectively.

Crop rotation and polyculture are traditional organic practices that also benefit Electroculture setups by enhancing soil health and biodiversity. Rotating different types of crops helps to prevent soil depletion, reduces the buildup of pests and diseases, and can improve soil structure and fertility. Planting a diversity of species can also enhance the ecological stability of the garden, reducing the need for chemical inputs and increasing resilience to pests and diseases.

Another sustainable practice is the use of natural pest control methods. These include introducing beneficial insects, using natural predators, or applying organic repellents and barriers. Such practices ensure that Electroculture setups do not harm beneficial insects and local wildlife, maintaining an ecological balance and supporting biodiversity.

Monitoring and adaptation are crucial components of sustainable Electroculture. Regularly testing soil conditions, monitoring plant health, and adjusting the Electroculture system accordingly helps in maintaining the effectiveness and environmental friendliness of the setup. This adaptive management approach ensures that Electroculture remains a sustainable part of the garden ecosystem, responding to changing conditions and needs without resorting to unsustainable practices.

Education and community engagement also play vital roles in promoting sustainable practices within Electroculture. Sharing knowledge and experiences about Electroculture and its benefits with the community can inspire others to adopt this innovative technology. Workshops, demonstration projects, and community gardens equipped with Electroculture systems can serve as powerful tools for education and outreach, spreading the principles of sustainability and organic farming.

By integrating these sustainable practices and organic solutions, Electroculture can be a highly effective and environmentally responsible technology for modern agriculture. It not only enhances plant growth and productivity but does so in a way that supports the health of the ecosystem, conserves resources, and promotes biodiversity, embodying the principles of sustainability in agriculture.

5

THE FUTURE OF ELECTROCULTURE

Current Research and Emerging Trends in Electroculture

Electroculture, a method of enhancing plant growth through electrical stimulation, has been garnering increasing interest and research in recent years. As our understanding of its principles deepens and technology advances, new developments and trends are emerging, promising to revolutionize the way Electroculture is applied in agriculture. This progress could significantly shape sustainable farming practices.

Recent research in Electroculture has focused on refining electrical stimulation techniques to maximize their efficacy and minimize potential harm to plants. Scientists are experimenting with variable frequencies, amplitudes, and pulse durations to determine the optimal electrical parameters for different species of plants. This tailored approach ensures that electrical stimulation promotes growth without stressing the plant tissues, leading to healthier plants and increased yields.

One of the most exciting trends in Electroculture is its integration with smart agriculture technologies. IoT (Internet of Things) devices, sensors, and automated systems are being combined with Electroculture to create more efficient and responsive farming systems. These technologies enable real-time monitoring and modifications of electrical inputs based on soil conditions, weather, and plant health indicators. This precision agriculture approach not only improves the effectiveness of Electroculture but also enhances water and energy conservation, which is crucial for sustainable farming practices.

As sustainability becomes increasingly important, the use of renewable energy sources in Electroculture systems is gaining traction. Solar-powered and wind-powered Electroculture systems are being developed to reduce the carbon footprint of agricultural electricity use. These systems harness natural energy sources to power the electrical stimulation of plants, aligning Electroculture with eco-friendly farming initiatives and reducing operational costs over time.

Researchers are exploring the application of Electroculture in hydroponics and aquaponics systems, where water serves as the growth medium for the plants. The conductivity of water can potentially enhance the distribution of electrical fields around the roots, potentially increasing the effectiveness of Electroculture in these systems. Preliminary studies have shown promising results, with significant improvements in plant growth rates and nutrient density, opening new avenues for integrating Electroculture into modern urban farming setups.

Advancements in the field of plant electrophysiology have provided deeper insights into how plants perceive and respond to electrical stimuli. This research is crucial for developing Electroculture techniques that work in harmony with natural plant processes. Understanding the electrophysiological responses of plants to electrical stimulation helps in designing systems that are both safe for plants and optimally effective for growth enhancement.

As interest in Electroculture grows, there is a significant push toward developing scalable systems that can be effectively used in larger agricultural operations. Researchers are working on ways to scale up Electroculture installations without compromising their efficiency or the uniformity of electrical field distribution. This involves innovations in system design, more durable materials, and better integration techniques that can be applied across various types of agricultural lands.

With the proliferation of Electroculture technologies, there is a growing need for educational programs to train farmers and agriculturists in the safe and effective use of these systems. Additionally, regulatory frameworks are being considered to ensure that Electroculture applications meet safety standards and are environmentally sustainable. These educational and regulatory efforts are crucial for the widespread adoption and responsible use of Electroculture in the agriculture industry.

The future of Electroculture looks promising, with ongoing research and emerging trends pointing towards more sophisticated, efficient, and sustainable applications. As this field continues to evolve, it holds the potential to significantly impact how food is grown, supporting global food security efforts while adhering to environmental sustainability principles.

Predictions and Innovations on the Horizon

The field of Electroculture is on the cusp of significant transformations, with emerging technologies and innovative research poised to reshape how electricity is used to enhance plant growth. As we look to the future, several predictions and potential innovations stand out, each promising to further integrate Electroculture into sustainable agricultural practices globally. These developments not only aim to increase the efficiency and effectiveness of Electroculture systems but also align them more closely with the principles of sustainability and environmental stewardship.

Future innovations in Electroculture are likely to focus on increasing the precision of electrical stimulation. With advancements in sensor technology and data analytics, Electroculture systems could soon automatically adjust electrical inputs based on real-time data regarding soil conditions, plant health, and environmental factors. This precision would minimize energy use while maximizing plant growth, leading to smarter, more resource-efficient farming.

Artificial intelligence (AI) has the potential to revolutionize Electroculture by optimizing the timing, intensity, and duration of electrical inputs in ways that humans cannot easily replicate. AI systems could analyze vast amounts of data from Electroculture operations to identify patterns and predict outcomes, leading to highly tailored treatment plans for individual crops or even specific plants. This could dramatically increase yield and quality while reducing the labor and resource costs associated with traditional farming methods.

As the demand for local and urban farming increases, there is a growing need for portable and modular Electroculture systems that can be easily implemented in a variety of settings, from small rooftop gardens to indoor agricultural facilities. These systems would be designed for easy installation and scalability, making Electroculture accessible to a broader range of users, including hobbyists and urban farmers.

Research is likely to expand into optimizing Electroculture for diverse climatic conditions, particularly extreme environments such as arid regions or areas with poor soil quality. By fine-tuning Electroculture systems to work under such conditions, it could be possible to significantly increase agricultural productivity in parts of the world where traditional farming is challenging or impossible. This would have profound implications for food security and economic development in these regions.

Future developments in Electroculture may include the use of biodegradable and eco-friendly materials for wires and other system components. Such materials would reduce the environmental impact of Electroculture installations and align with broader goals of sustainability. Additionally, research into less invasive and more natural forms of electrical conduction, such as using plant-based conductors, could further decrease the ecological footprint of these systems.

There may be increased collaboration between Electroculture research and plant breeding programs. By understanding how certain plant varieties respond differently to electrical stimulation, breeders could develop new varieties that are optimized for Electroculture. This combination could result in plants that grow quicker, produce more, and are more resistant to environmental challenges, all while requiring less input from other sources.

As Electroculture becomes more prevalent, there will likely be a push towards developing regulatory frameworks and standards to ensure the safety, efficacy, and environmental compatibility of Electroculture systems. This standardization would help facilitate wider adoption and ensure that Electroculture technologies are used responsibly and sustainably.

The future of Electroculture in sustainable agriculture looks promising, with numerous innovations on the horizon that could further enhance its application and effectiveness. These advancements could lead to more sustainable agricultural practices globally, offering solutions to some of today's most pressing challenges, including food security, resource conservation, and environmental protection. As research continues and technology evolves, Electroculture is set to play a pivotal role in shaping the future of agriculture.

CONCLUSION

Reflecting on Your Electroculture Journey

Embarking on an Electroculture journey offers a unique blend of challenges and rewards, providing an opportunity for both personal and professional growth within the realm of sustainable agriculture. Reflecting on this journey, one can appreciate the key learnings and the myriad ways in which Electroculture enhances our understanding of plant biology, environmental stewardship, and agricultural innovation.

Electroculture is not merely about applying electrical currents to stimulate plant growth; it's an exploration of the subtle interactions between electricity and biological systems. Through this process, practitioners gain a deep appreciation for the complexities of plant physiology. Learning how plants respond to electrical stimuli can open up new perspectives on how to care for and nurture crops more effectively, leading to healthier plants and potentially larger yields.

Moreover, engaging with Electroculture can significantly broaden one's skills in both the technical and environmental aspects of agriculture. It involves a meticulous blend of electrical engineering and soil science, demanding a precise understanding of how electrical systems interact with different soil types, plant species, and environmental conditions. This technical proficiency is paired with a growing awareness of sustainable practices. Electroculture encourages the minimization of chemical inputs, such as fertilizers and pesticides, promoting a more organic approach to farming that is kinder to the environment and often more cost-effective in the long run.

One of the most significant personal growth opportunities presented by Electroculture is the development of problem-solving and innovative thinking skills. Setting up and maintaining an Electroculture system requires a proactive approach to troubleshooting and innovation. Practitioners learn to identify the signs of system inefficiencies or failures quickly and must think creatively to solve these issues, whether they involve adjusting electrical outputs, modifying system layouts, or experimenting with different soil amendments to improve conductivity and plant health.

Electroculture also offers an excellent opportunity for community engagement and education. Those who delve into this field often find themselves at the forefront of agricultural technology, positioned to educate others about the benefits and methods of Electroculture. This can lead to community projects, educational workshops, and collaborations with other farmers and researchers, expanding one's network and influencing the wider adoption of sustainable agricultural practices.

Finally, the journey through Electroculture often leads to a greater sense of responsibility towards global challenges such as food security and climate change. By optimizing the use of resources and reducing dependency on chemical inputs, Electroculture practitioners can contribute to a more sustainable future in agriculture, making a positive impact on the environment and society.

The journey through Electroculture is rich with learning opportunities and personal growth. From enhancing technical and environmental knowledge to fostering community involvement and addressing global challenges, Electroculture offers a profound way to engage with the world of agriculture. Reflecting on these experiences, practitioners can appreciate the extensive skills they've developed and the positive impact they've made, both in their local communities and in the broader agricultural context.

The Future of Electroculture Gardening

The future of Electroculture gardening is bright, with tremendous potential for growth and innovation. As we look ahead, it's clear that continuous exploration and widespread adoption of Electroculture techniques could play a pivotal role in transforming agricultural practices worldwide. Encouraging this progression will not only lead to more efficient and sustainable farming but also foster a deeper connection between technology and traditional cultivation methods.

Electroculture, which harnesses the power of electrical stimulation to enhance plant growth, is still in its nascent stages when compared to more traditional agricultural practices. Yet, its promise is evident, offering a way to increase crop yields, improve plant health, and reduce the reliance on chemical inputs such as fertilizers and pesticides. Modern farmers and gardeners are finding Electroculture to be an increasingly appealing alternative because to its efficiency and environmental friendliness, particularly as the world's population continues to rise and environmental issues gain greater urgency.

To ensure the future success and expansion of Electroculture gardening, there are several key areas where continued effort and focus will be essential:

Research and Development

Investing in research is crucial for uncovering the full potential of Electroculture. This involves not only refining existing techniques but also exploring new applications and innovations. Research can help optimize the parameters used in Electroculture systems, such as voltage levels, frequency, and timing, to suit different types of plants and environmental conditions. Additionally, understanding the deeper biological impacts of electrical stimulation on plants can lead to better and more targeted applications of the technology.

Education and Training

For Electroculture to gain broader acceptance and usage, educating farmers and gardeners about its benefits and methodologies is vital. This can be achieved through workshops, courses, and demonstrations that not only teach the practical aspects of setting up and running Electroculture systems but also highlight the science behind the technique. Educating the next generation of agriculturists at universities and agricultural colleges about Electroculture could also inspire young innovators to develop and champion this technology further.

Technology Integration

Integrating Electroculture with other agricultural technologies, such as automated irrigation systems, precision farming tools, and artificial intelligence, can enhance its efficiency and effectiveness. For example, using data-driven insights from IoT sensors to dynamically adjust the electrical inputs in an Electroculture system could optimize plant growth conditions without human intervention, making the process more sustainable and less labor-intensive.

Community and Collaborative Efforts

Creating an Electroculture community of practice can help to exchange best practices, experiences, and information. Collaborations between researchers, technology developers, and end-users are essential to drive innovation and solve practical challenges in the field. Community gardens and local farming cooperatives adopting Electroculture can also demonstrate its benefits to a wider audience, helping to normalize and boost confidence in this approach.

Regulatory Support and Incentives

For Electroculture to be adopted on a larger scale, supportive policies and incentives from governments and international bodies are necessary. Regulations that recognize and promote sustainable agricultural innovations can accelerate the adoption of Electroculture. More widespread use could also be encouraged by financial incentives, such as subsidies or tax exemptions for farmers who use sustainable technology like Electroculture.

Global Outreach

Expanding the reach of Electroculture to developing regions where agricultural innovation can have a profound impact on food security and economic stability is another important goal. Tailoring Electroculture systems to fit the specific needs and conditions of these regions will be crucial to their success and acceptance.

The future of Electroculture gardening holds the promise of revolutionizing how we grow food, making agriculture more sustainable, efficient, and productive. By continuing to explore, adopt, and integrate Electroculture techniques, we can ensure that this innovative approach has a lasting and positive impact on global agriculture.

Encouragement and Final Tips

As we reflect on the transformative potential of Electroculture, it's clear that this innovative approach to agriculture offers more than just a method to enhance plant growth; it represents a shift towards more sustainable and efficient farming practices. Whether you are a seasoned gardener, a commercial farmer, or a new enthusiast, embracing Electroculture can lead to significant advancements in how we cultivate our crops and interact with our environment. Here are some final words of encouragement and tips to inspire continued use, experimentation, and community involvement in Electroculture.

Firstly, remain curious and open to experimentation. There are several potential to innovate and enhance current procedures in the ever-evolving field of Electroculture. By experimenting with different electrical parameters, plant types, and setups, you can discover optimizations that yield even better results for your specific environment and crop choices. Document your processes and outcomes to build a body of knowledge that can benefit others while advancing your own understanding and expertise.

Secondly, embrace the community aspect of Electroculture. Joining forums, attending workshops, and participating in local gardening clubs can connect you with like-minded individuals who share your interest in sustainable agriculture. These communities serve as invaluable resources for support, learning, and collaboration. Sharing your experiences and learning from others not only enriches your own Electroculture practice but also helps to foster a broader acceptance and understanding of this technology.

Don't hesitate to reach out to academic and research institutions that are exploring Electroculture. Collaborating on research projects can provide access to cutting-edge knowledge and resources that can enhance your own practices. This collaboration also bridges the gap between practical and theoretical knowledge, pushing the envelope of what's possible in Electroculture.

Advocate for and educate others about the benefits of Electroculture. There's a significant opportunity to promote Electroculture as a key element of sustainable agricultural practices. By conducting demonstrations, writing articles, or even hosting informational sessions, you can raise awareness and interest in Electroculture, potentially inspiring new adopters and advocates for this innovative approach.

Finally, never underestimate the impact of your contributions, no matter how small they may seem. Each individual who adopts and promotes Electroculture is a vital part of the journey towards more sustainable and efficient farming practices worldwide. Your efforts contribute not only to your own success but also to the global mission of creating a more sustainable future for agriculture.

In closing, let your journey with Electroculture be guided by a spirit of innovation, community, and sustainability. Continue to explore, experiment, and share your findings. The path of Electroculture is not just about enhancing plant growth; it's about cultivating a deeper understanding and respect for our natural world. Let's move forward with the knowledge that each small step we take is part of a much larger leap toward sustainable living and environmental stewardship. Keep growing, keep experimenting, and keep sharing your journey with the world. Your work is vital, your discoveries valuable, and your potential to inspire change is immense.

Thank You for Growing with Us!

We sincerely hope that **Electroculture Gardening** has provided you with valuable insights and practical tips to enhance your gardening practices. If this book has been beneficial to you, please take a few moments to leave us a review on Amazon. Your feedback helps us to improve and informs fellow gardeners about the practical applications of electrophilic culture.

How You Can Share Your Review:

Through Amazon.com:

- Go to the Amazon page where you found my book.
- Navigate to the 'Customer Reviews' section.
- Click on 'Write a customer review' to share your valuable insights.
- Instant QR Code Access: Simply scan the QR code below with your smartphone to be directed to the Amazon review section.

www.ingramcontent.com/pod-product-compliance
Lightning Source LLC
Chambersburg PA
CBHW082241220526
45479CB00005B/1297